INTRODUCTION TO
PERCOLATION THEORY

INTRODUCTION TO PERCOLATION THEORY

Dietrich STAUFFER

Institute of Theoretical Physics, Cologne University, FR Germany

Taylor & Francis
London and Philadelphia
1985

UK   Taylor & Francis Ltd, 4 John St., London WC1N 2ET

USA   Taylor & Francis Inc., 242 Cherry St., Philadelphia,
PA 19106–1906

British Library Cataloguing in Publication Data

Stauffer, Dietrich
   Introduction to percolation theory.
   1. Percolation—(Statistical physics)
   I. Title
   532'.7   QC174.85.P45

   ISBN 0-85066-315-6

Library of Congress Cataloging in Publication Data

Stauffer, Dietrich.
   Introduction to percolation theory.

   Includes index.
   1. Percolation (Statistical physics)
I. Title.
QC174.85.P45S73   1985   530.1'3   85-4793
ISBN 0-85066-315-6

*Filmset by Mid-County Press, London SW15 2NW
Printed in Great Britain by Taylor & Francis (Printers) Ltd,
Basingstoke*

CONTENTS

A table of percolation thresholds is printed on page 17, a table of critical exponents on page 52

PREFACE

This book is an attempt to introduce the reader to a research field which is already more than forty years old but which has become very fashionable in the research publications of the last ten years. More than a hundred publications are printed each year where 'percolation' or similar words appear in the title. But in contrast to many other modern research fronts, percolation theory is a problem which is, in principle, easy to define. It is, however, not so easy to solve. Thus percolation theory gives the reader the opportunity to enter current research without having to hear many specialized courses or to read voluminous textbooks. Percolation theory has been taught to first year undergraduates and even younger students by this author and by others, but it has also been utilized in courses for more advanced students on disordered systems or on computer simulations. The present book tries to be useful for all these purposes. Since percolation is not, and in my opinion should not be, a standard subject for university examinations, it is addressed mainly to readers interested in research. But it is aimed at one who is just starting research in this field, not to readers who have already worked in percolation theory. These experts will doubtless notice my biased selection of material.

The prerequisites are familiarity with computer programming (e.g. Fortran), integration and differentiation of functions of one variable, and probabilistic elements like averages or statistical independence. Therefore the book should be understandable to students in computer science, mathematics, chemistry or biology who might be interested in stretching a big computer to its limits by efficient programming, in simple applications of probability theory, in theories for the gelation of branched macromolecules, or in the spread of epidemics in an ensemble of living beings. But presumably the book will be read mostly by physics students since the methods by which it attacks the percolation problem are taken from the theory of phase transitions like ferromagnetic Curie points. For these we will mention throughout the book analogies between the geometrical aspects of percolation and the physical aspects of thermal phase transitions. Any modern textbook on Statistical Physics will give the background to understand these analogies, and students not interested in the analogies may simply ignore them.

Some readers may be acquainted with the scaling theory of phase transitions developed during the last twenty years and honoured by the 1982 Nobel prize for Kenneth G. Wilson. They will notice that many aspects of percolation theory are simply borrowed from the physics of phase transitions, as will be mentioned in the text. On the other hand, a reader who has not yet learned scaling theory or renormalization group methods for general phase transitions, but who wants to know something about them, may use percolation theory as a starting point. In many respects percolation is the simplest not exactly solved phase transition and thus may serve as an introduction to the sometimes more difficult review articles or books on phase transitions and critical phenomena.

What this book does not try to be is mathematically rigorous or complete in dealing with the actual state of research. We only give cursory mention to the applications of percolation since they require a more specialized readership. We will try however to list in the literature some of the more thorough review articles on percolation for those who want to study this field further. However, because of the rapid development of percolation, the reader should not assume that these references are still the most recent relevant reviews or original articles at the time he reads this book.

I am indebted to J. Kertész for information about his percolation seminar at Munich Technical University, and his comments and those of D.W. Heermann, H. J. Herrmann, A. Margolina, B. Mühlschlegel, R. B. Pandey, S. Redner, and M. Sahimi on a preliminary version of the manuscript (though I did not follow all of their suggestions, like calling this book 'My biased view of percolation'), M. Suessenbach for producing Figure 2, and A. Schneider for drawing the other figures. The manuscript was written using the text editing system of a PDP 11/34 computer, thanks to the efforts of A. Weinkauf and M. Schulte; needless to say this computer should be blamed for all the errors in the book.

Dietrich Stauffer
January 1985

CHAPTER 1

introduction: forest fires and diffusion

1.1. What is percolation?

Imagine a large array of squares as shown in Figure 1(a). We imagine this array to be so large that any effects from its boundaries are negligible. Physicists call such an array a square lattice, mathematicians denote it by $\mathbb{Z}^2$; common sense identifies it with a big sheet of ruled paper. (You may complain that the square lattice in Figure 1(a) is not very large, but the publisher did not allow me to fill all remaining pages of this book with these squares, which would have greatly simplified my task of writing the book and yours of reading it.) Now a certain fraction of squares are filled with a big dot in the centre, whereas the other squares are left empty, as in Figure 1(b). We now define a *cluster* as a group of neighbour squares occupied by these big dots; these clusters are encircled in Figure 1(c). From this picture we see that squares are called neighbours if they have one side in common but not if they only touch at one corner. Physicists call squares with one common side 'nearest neighbour sites on the square lattice', whereas squares touching at one corner only are 'next nearest neighbours'. All sites within one cluster are thus connected to each other by one unbroken chain of nearest-neighbour links from one occupied square to a neighbour square also occupied by a big dot. The graphical 'cluster' explanation through Figure 1(c) seems more appropriate for our purposes here than a precise mathematical definition. Percolation theory now deals with the number and properties of these clusters; perhaps the reader will agree with me that there are not many requisites needed to understand what percolation theory is about.

How are the dots distributed among the squares in Figure 1? One may assume that the dots love to cling together, or that they hate each other and try to move as far away from each other as possible. But the simplest assumption is that they ignore each other, not unlike scientists working in similar fields. Then the occupation of the squares is *random*, that is each square is occupied or empty independent of the occupation status of its neighbours. We call p the probability of a site being occupied by a big dot; that means that if we have N squares, and N is a very large number, then pN of these squares are occupied, and the remaining

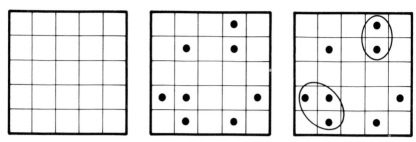

Figure 1. Definition of percolation and its clusters. Figure 1(a) shows parts of a square lattice, in Figure 1(b) some squares are occupied with big dots, in Figure 1(c) the 'clusters', groups of neighbouring occupied squares, are encircled except when the 'cluster' consists of a single square.

$(1 - p)N$ of these squares are empty. This case of random percolation is what we concentrate on here:

Each site of a very large lattice is occupied randomly with probability p, independent of its neighbours. Percolation theory deals with the clusters thus formed, in other words the groups of neighbouring occupied sites.

Of course, the reader may replace 'occupied by a big dot' with 'black' and 'empty' with 'white' (or 'red', if he likes politics); or he may use any other suitable pair of words denoting two mutually exclusive states of the site.

Figure 2 shows a computer-generated sample of a 60×60 square lattice, with probability p increasing from 10% to 90%. We see that for p above 0·6 one cluster extends from top to bottom and from left to right of the sample; one says that this cluster percolates through the system rather like water percolates through a coffee machine. A large part of this book deals with the peculiar phenomena of percolation near that concentration p_c where for the first time a percolating cluster is formed. These aspects are called critical phenomena, and the theory attempting to describe them is the scaling theory.

Historically percolation theory goes back to Flory and to Stockmayer who during World War II used it to describe how small branching molecules form larger and larger macromolecules if more and more chemical bonds are formed between the original molecules. This polymerization process may lead to gelation, that is to the formation of a network of chemical bonds spanning the whole system. Thus the original small molecules correspond to our squares, the macromolecules to our clusters, and the network to our percolating cluster. You may be an experienced researcher in percolation without having been aware of it, for the boiling of an egg, which is first liquid and then becomes more solid-like ('gel') upon heating is an example. Flory and Stockmayer developed a theory which today one calls percolation theory on the Bethe lattice (or Cayley tree) and which will be explained later. But even today it is controversial whether critical phenomena for gelation are described correctly by percolation theory and its assumption that chemical bonds are formed randomly (de Gennes (1976); Stauffer, Coniglio and Adam (1982)).

Usually, the start of percolation theory is associated with a 1957 publication of Broadbent and Hammersley which introduced the name and dealt with it more mathematically, using the geometrical and probabilistic concepts explained above. Hammersley, in his personal history of percolation in *Percolation Structures and Processes*, mentions that the new computers which became available to scientists at that time were one of the reasons for developing percolation theory as a problem where the computers could be useful. We will see later that even today computers play a crucial role for percolation, with lattices containing thousands of millions of sites being simulated and analysed.

The percolation theory as described here, with its particular emphasis on critical phenomena, was developed since the 1970s; one may regard a note by Essam and Gwilym in 1971 as one of the starting points of the later avalanche of publications. Instead of going through the details now we describe two simple 'games' which can be easily simulated on a computer and which may serve as an introduction to a reader preferring to learn percolation by a 'hands-on' approach. These examples are somewhat unusual, and the reader may skip them and proceed with Chapter 2.

1.2. Forest fires

This section introduces a simple model for forest fires. Its aim is not so much to help fighting fires but to help to understand the idea of a percolation threshold, the concept of a sharp transition with diverging times, and computer simulation.

French scientists in Marseilles and elsewhere are interested, for obvious reasons, in understanding and controlling forest fires. They told me of the following percolation problem which can easily be simulated on a computer. How long does a forest fire take to either penetrate the forest or to be extinguished?

As is well known, a diligent student should make hundreds of independent experiments to reduce statistical errors before reporting the result in his thesis. If for every thesis, a hundred fires were initiated in the forests surrounding the university, society's respect for research might be diminished. It is much more practical to simulate numerous such fires on a computer. For this purpose we approximate the forest by a square lattice. Each square in Figure 1 is either occupied by a tree, in which case we call that site 'green', or it is empty, in which case we call it 'white'. The probability for a green square is p, that for a white square is $(1 - p)$. For $p = 1$ all squares would correspond to trees, which would be appropriate to a garden of apple trees but not for a natural forest. The fact that $p < 1$ allows for holes (white squares) which cause disorder in the forest. This distribution of white and green sites (squares) is our initial state.

$P = 0.10$

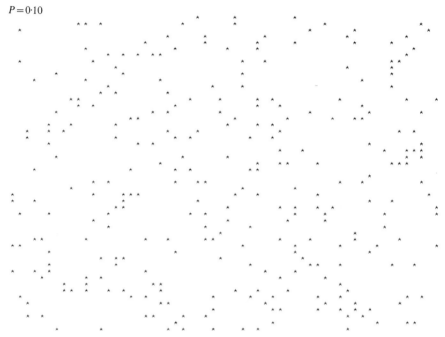

Figure 2. Example for percolation on a 60 × 60 square lattice, for various p as shown in the figure. Occupied squares are shown as *, empty squares are ignored. Near the threshold concentration 0·5928 we have marked the largest cluster.

Now we let some trees burn and call those squares which correspond to burning trees 'red' sites. The simplest choice is to light all the trees in the first row of the lattice, whereas the remaining trees, in lines 2, 3, ..., L of the $L \times L$ lattice, remain green. Does this fire on one side of the forest penetrate through the whole forest down to line L of our array?

For this purpose we have to clarify how a tree can ignite the other trees. To simplify the computer simulation we go through our lattice regularly, first scanning the first line of trees from left to right and checking which neighbours they ignite, then scanning the second line in the same way, and so on until we reach the last line of trees. During the whole simulation, a green tree is ignited and becomes red if it neighbours another red tree which at that time is still burning. Thus a just-ignited tree ignites its right and bottom neighbour within the same sweep through the lattice, its top and left neighbour tree at the next sweep. Reaching the end, we start again with the tree at the extreme left in the first line. Each sweep through the whole lattice (experts call that one Monte Carlo step per site) constitutes one time unit in our simulation. We assume that the fire can spread only to green nearest neighbour trees, not to trees which are further away. Furthermore, a tree which has burnt during one time unit is regarded as burnt out ('black') and no longer ignites any other tree. We regard the forest fire as

$P = 0.20$

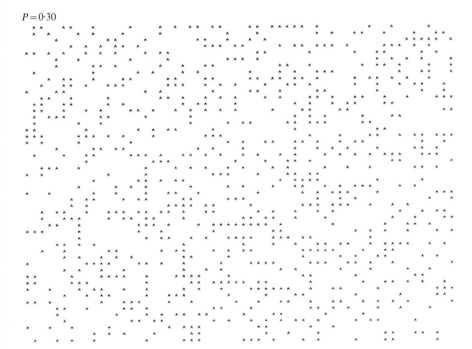

$P = 0.30$

$P = 0.40$

$P = 0.50$

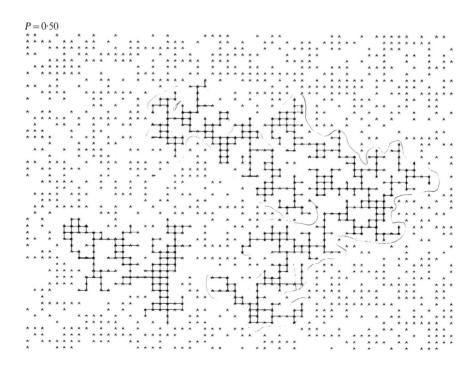

P = 0·60

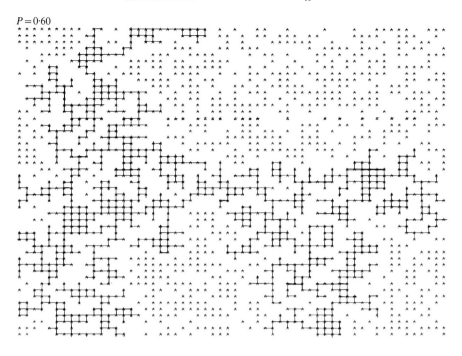

P = 0·70

$P = 0.80$

$P = 0.90$

terminated if it either has reached the last line or if no burning trees are left. (In the first case, the fire would ignite the next line of trees if a larger lattice had been stored in the computer; in the second case, only black trees and green trees adjoining white places are left over, the black trees constituting formerly burning trees which have burnt out, the green trees never having been touched by the fire since they were separated safely from the burning trees). The lifetime of the forest fire is defined as the number of sweeps through the lattice until termination is reached, averaged over many distributions of trees among the sites of the same lattice at the same probability p.

Figure 3 shows this lifetime of forest fires as a function of the probability p that a square is occupied with a tree. These simple computer simulations indicate that there is a sharp transition, for the above case near $p = 0.6$, where the lifetime seems to approach infinity. Of course, in the simulation of finite lattices the reader cannot expect truly infinite times; but one can simulate the forest fires at the same 'critical' value of p near 0.5928 for different lattice sizes and show that the lifetime increases with increasing size of the forest.

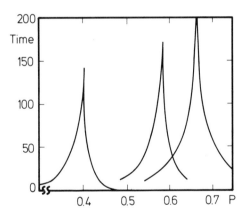

Figure 3. Average termination time for forest fires, as simulated on a square lattice. The centre curve corresponds to the simplest case described first in the text. The lefthand curve gives data if the fire can spread to both nearest and next-nearest neighbours. For the righthand curve two burning trees are needed to ignite a nearest or next-nearest neighbour.

Why is there a special value of p, which we call the percolation threshold p_c, where the lifetime seems to diverge? For p near unity, each row can immediately ignite the trees in the next row, and thus after one sweep through the lattice the fire may already have reached the last row. For p near zero, most burning trees have no neighbours at all, and the fire stops there after the tree has burnt out; thus after a few sweeps nothing burns anymore. If we increase p from small values to large values, then at some critical value $p = p_c$ a path of neighbouring trees appears which connects the top row with the bottom row for the first time, that is we see a percolating cluster. The shortest path which, for p slightly above p_c, this

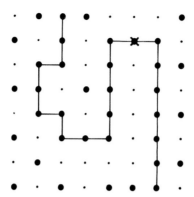

Figure 4. Example of the shortest path connecting the top line of a small square lattice with the bottom line, for p slightly above p_c. The straight sections of this line connect the centres of occupied squares. The X marks a site which, if missing due to a small reduction of p, would disconnect top and bottom lines but would still give a long termination time for the forest fire simulation. ·

percolating network creates to connect top and bottom will in general be very different from a straight line. Figure 4 shows a typical path.

Because of the simplified way in which we constructed our model the fire spreads preferably from top to bottom, or left to right, and needs a much longer time to move backwards from right to left or from bottom to top. For four consecutive forward steps, say top to bottom, it needs just one time unit, whereas four backward steps require four time units, as the reader can easily check on this figure by going through the above algorithm. Thus now the fire needs a long time to penetrate the forest. If p is diminished to a value slightly below p_c, then some trees, for example the one marked by an X in our Figure 4, may be missing. The fire then needs a long time to find out that it cannot penetrate the forest, and thus only after many sweeps through the lattice will the fire be extinguished. Therefore the lifetime will become very large if p approaches p_c from below or above.

We also show in Figure 3 the results for two modifications of the above model. In one case we allow the fire to spread not only to the nearest neighbour trees (squares which have one side in common) but also to next-nearest neighbours (squares which have only one corner in common). Then the critical point is shifted to about 0·4; experts have shown that it is at one minus the above critical value, i.e. $1 - 0·5928 = 0·4072$. But even without much thinking and computing one can understand that now the fire can spread more easily since it can jump over longer distances; therefore the percolation threshold is lowered.

The other modification goes in the opposite direction: We assume that the weather is more like in Nova Scotia (Canada) than in Marseilles (France). Since it is quite cold, a tree needs two burning neighbours, instead of only one, before it can ignite. Now it is more difficult for the fire to percolate through the forest, and the percolation threshold is shifted upwards, as the simulations in Figure 3 show.

The reader may complain that the above algorithm gives the fire a preference to spread to the right and the bottom and may dislike these similarities with political or economic trends, respectively. But for forest fires, such trends can be justified as representing a wind blowing in one 'diagonal' direction. In reality this preference is introduced to save computer time. (One sees here how the species Homo Veritabile Sapiens, also known as the Theoretical Physicist, is working: one introduces a model because one can work with it. If Nature disagrees with the model, Nature will be denied a degree in Theoretical Physics. After some time, Nature usually reapplies for the examination since a real system is found corresponding roughly to the theoretical physicist's model. Student exercise: Guess the author's profession.) This author has neither the mental nor the hardware capability to let the fire burn on a colour computer monitor.

For readers interested in the physics of phase transitions it should be mentioned that the percolation threshold at $p = p_c$ gives the position of a phase transition (for experts only: without 'broken symmetry'). At a phase transition, a system changes its behaviour qualitatively for one particular value of a continuously varying parameter. In the percolation case, if p increases smoothly from zero to unity, then we have no percolating cluster for $p < p_c$ and (at least) one percolating cluster for $p > p_c$. Thus at $p = p_c$, and only there, something peculiar happens: for the first time a path of neighbouring green trees connects top and bottom. Also the divergence of characteristic times (in our case the fire lifetime) at the critical point has analogies in other phase transitions where it is called 'critical slowing down'. For example, for a temperature only slightly below the liquid–gas critical temperature, the fluid is quite unsure whether it wants to be liquid or vapour, and thus takes a lot of time to make its choice; this time can be measured by light scattering. Similarly, relaxation times near magnetic Curie points are very large.

1.3. Diffusion in disordered media

Hydrogen atoms are known to diffuse through many solids, an effect which might become important for energy storage. If the solid is not a regular lattice, this diffusion takes place in a disordered, not an ordered medium. A particularly simple disordered medium is our percolation lattice, where only a fraction p of all sites (squares) is occupied, the rest are empty. Let us assume the hydrogen atom can move only from one occupied site of the lattice to a nearest neighbour which is also occupied. Then the motion is restricted to the cluster of percolation theory to which the atom belongs initially. It can never jump to another cluster since then it would have to move at least once over a distance larger than that between nearest neighbours. This problem was called the 'ant in the labyrinth' by de Gennes in 1976; and Mitescu and Roussenq then made the first computer simulations following his suggestion. At the beginning of the 1980s this problem became very fashionable, particularly at the percolation threshold $p = p_c$.

Let us not care whether hydrogen atoms move through solids, an ant tries to escape a labyrinth, or the reader desperately searches for a way through this book. We simply have a point, called an ant, which sits on an occupied square of our square lattice and which at every time unit makes one attempt to move. This attempt consists in randomly selecting one of its four neighbour squares. If that square is occupied, it moves to that square; if instead it is empty, the ant stays at its old place. In both cases the time t is increased by one unit after the attempt. After a certain time t, one calculates the squared distance between the starting point and the end point. One repeats the simulation by giving the ant a different occupied square as a starting point; finally, one averages the squared distance obtained in this way over many ant movements on many lattices at the same p and same lattice size. How does R, the square root of this averaged squared distance (also called the root mean square or rms displacement) depend on time t?

For $p = 1$ one has diffusion on a regular lattice without disorder, and elementary statistical considerations give $R^2 = t$ exactly, if our squares have a length equal to unity. (Proof: For each such random walk, the end-to-end vector R is the vector sum of t displacement vectors d_i, $i = 1, 2, ..., t$. When we calculate the square of that sum and then its average, we have to calculate the averages of the scalar products $d_i d_j$. For $i = j$, this scalar product is simply the square of the nearest neighbour distance, which is unity. For i and j different, the scalar product can be $+1$ or -1 with equal probability since we assumed that the motion is completely random. Moreover, in half of the cases the scalar product is zero since d_i and d_j are perpendicular to each other. Thus on average this product cancels out except for $i = j$ where it gives unity. Therefore the squared sum equals t. This proof is not necessary to understand the remainder of the book since we will mainly deal with problems which are not exactly solved.)

Figure 5 shows the results of simple computer simulations on the square lattice. On this double-logarithmic plot one sees the power law $R = const\; t^k$ more

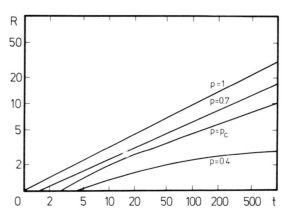

Figure 5. Example of the distance R travelled by an ant in a labyrinth, as simulated on a square lattice for $p < p_c$, $p = p_c = 0.5928$, and $p > p_c$. Note the double-logarithmic scales.

easily than on a normal diagram. It seems to describe the relation between distance R and time t for sufficiently large t. Since $\log R = \log (const) + k \log (t)$, power laws show up as straight lines in such log-log plots, with the slope giving the exponent k of the power law. We see that for a concentration p far above $p_c = 0.59$ k is near 1/2 for large times, whereas for p far below p_c the distance R approaches a constant for large times, that is $k = 0$. Right at the critical point k takes on a value in between these two extremes and is roughly equal to 1/3. This effect, that the exponent is neither that for normal diffusion ($k = 1/2$) nor that for a constant distance ($k = 0$) was called anomalous diffusion by Gefen, Aharony and Alexander. Again it has an analogue in critical points near thermal phase transitions. For example, spin diffusion in ferromagnets at $T = T_c$ or mass diffusion at liquid–gas critical points no longer follows the normal diffusion laws but is often described by an anomalous diffusion exponent k or $1/z$.

It is easy to understand why the ant moves so differently for p above and below the percolation threshold p_c. For $p < p_c$ there are only finite clusters present, and the ant sits on one of them. Thus it moves only within that cluster (if this cluster happens to be an isolated occupied square, the ant cannot move at all.) Therefore its motion is restricted over finite distances, and R approaches a value connected to the cluster radius if t is very large. For $p > p_c$, on the other hand, the particle can move to infinity if it starts on the percolating network. There are certain holes in this network; but for distances larger than the typical hole size, the ant feels only an average over the small holes, just as the tyres of your Rolls-Royce average over the small pores of the asphalt over which your chauffeur is driving you. Thus the disorder acts as a friction which slows down the diffusion process but does not prevent it: $k = 1/2$ for long times. Only at the border case $p = p_c$, does the ant not know which of the two power laws it should follow; if the reader wants to know he has to wait until Chapter 5.

(Only for experts in phase transition theory we mention here that a lot of models investigated during the last decades are special cases of the so-called n vector model, as far as their critical behaviour is concerned. The integer n equals unity for the lattice gas or Ising model, two for superfluid helium, three for Heisenberg magnets, infinity for the spherical model or Bose–Einstein condensation, zero for self-avoiding walks (polymer chains with excluded volume effect), and minus two for random walks (polymer chains without excluded volume effects). The Ising model, on the other hand, is also the case $q = 2$ of the so-called q-state Potts model. And percolation corresponds to $q \rightarrow 1$ in this Potts model. Therefore it is not surprising that all these phase transitions have many similarities with each other and with percolation.)

Further reading

General reviews are collected in the book *Percolation Structures and Processes*, edited by G. Deutscher, R. Zallen and J. Adler, published in 1983 by Adam Hilger, Bristol, as

Annals of the Israel Physical Society 5. Out of the 21 articles there, our introduction is related in particular to that of Hammersley on the origins of percolation theory, of Jouhier *et al.* on gelation of macromolecules, and of Mitescu and Roussenq on ant diffusion.

Mathematical aspects, mostly ignored by us, are emphasized by H. Kesten's book *Percolation Theory for Mathematicians*, published in 1982 by Birkhauser, Boston.

Some other percolation reviews in regular journals are:
Clerc, J.P. *et al.*, *Annales de Physique (Paris)*, **8**, 1, (1983).
de Gennes, P.G., *La Recherche*, **7**, 919, (1976).
Domb, C., Stoll, E. and Schneider, T., *Contemporary Physics*, **21**, 577, (1980). A percolation movie in colours but without sound.
Essam, J.W., *Reports on Progress in Physics*, **43**, 843, (1980).
Kirkpatrick, S., *Reviews of Modern Physics*, **45**, 574, (1973).
Sahimi, M., in *Percolation, Random Walks, Modeling and Simulation*, Lecture notes in Mathematics 1035, p. 314 (Heidelberg: Springer Verlag, 1983). A summary with emphasis on numerical simulations for critical phenomena.
Shante, V.K.S. and Kirkpatrick, S., *Advances in Physics*, **20**, 325, (1971).
Stauffer, D., Coniglio, A. and Adam, M., *Advances in Polymer Science*, **44**, 103, (1982).

For phase transitions in general see:
Stanley, H.E., *Introduction to Phase Transitions and Critical Phenomena*, (Oxford: OUP, 1971).
or the continuing series of books
Phase Transitions and Critical Phenomena, edited by C. Domb and M.S. Green, then by C. Domb and J.L. Lebowitz, (New York: Academic Press, 1972–0000).

For time dependent aspects near phase transitions see:
Hohenberg, P.C. and Halperin, B.I., *Reviews of Modern Physics*, **49**, 435, (1977).

For forest fires in particular see:
Mackay, G. and Jan, N., *J. Phys. A.*, **17**, L757.

For percolation in semiconductors see:
Shklovskii, B.I. and Efros, A.L., *Electronic Properties of Doped Semiconductors*, (Heidelberg: Springer Verlag, 1984).

CHAPTER 2

cluster numbers

2.1. The truth about percolation

I did not tell you the whole truth: life is more than just a square lattice. There are also the triangular lattice, the honeycomb lattice, and other two-dimensional structures. In three dimensions we have the simple cubic lattice, the body centred cubic (bcc) lattice, the face centred cubic (fcc) lattice, the diamond lattice, among others. Dimensions higher than three are also useful to test theories, and usually are treated by the hypercubic lattice. In Figure 1 we defined the square lattice through the centres of the squares shown there. We could also have defined it equivalently through the points where the lines in Figure 1 cross.

Now in Figure 6(a) the situation is different. When we put the sites of the lattice on the crossing points of the lines of Figure 6(a) we obtain the triangular lattice; if instead we put them in the centres of the triangles with equal distance from the surrounding lines, we get the honeycomb lattice. (I do not recommend the use of the term hexagonal lattice.) Figure 6(b) consists of cubes and is called $\mathbb{Z}^3$ by mathematicians; it does not matter whether we put the lattice sites in the centres or on the corners of the cube. In a Fortran computer program one could store the sites of the simple cubic lattice in an array $A(i, j, k)$ whose indices i, j and k vary independently from 1 to L, where L is a large integer. The sites of the bcc lattice are both the corners and the centres of the cubes, whereas the fcc lattice consists of the centres of the cubes and the centres of the six faces surrounding each cube. Diamond lattices are not the programmer's best friends. A five-dimensional hypercubic lattice is much easier to program, for example by using a Fortran array $A(i, j, k, m, n)$ with five independent natural numbers as indices (or different Fortran statements to the same effect) to simulate a part of $\mathbb{Z}^5$.

To demonstrate percolation experimentally on a triangular lattice we may put numerous small spheres of equal size but with two different colours (black and white, for example) into a large box. These balls will roll around on the bottom of the box if the box is large enough in order to prevent the spheres being on top of each other. Now if the box is slightly inclined all the balls will roll to one side of the bottom plane. By shaking the box slightly, the balls are persuaded to

15

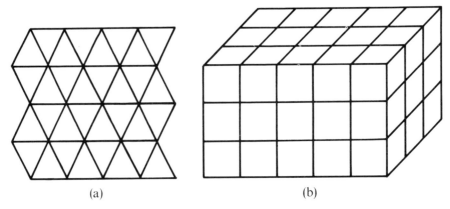

(a)   (b)

Figure 6. Definition of triangular, honeycomb and cubic lattices. For the triangular lattice, every intersection of the lines in Figure 6(a) is a lattice site; for the honeycomb lattice, the centres of the triangles in Figure 6(a) form the lattice sites. The simple cubic lattice consists of the corners of the cubes in Figure 6(b); for the bcc lattice the centres of the cubes, for the fcc lattice the centres of the six faces for each cube, are added to the simple cubic lattice.

form a triangular lattice with a few defects. The two different colours then symbolize occupied and empty sites, and one sees the clusters quite directly. With an equal number of black and white balls, one is studying the behaviour at the percolation threshold $p_c = 1/2$ (see below). If the black balls are electrically conducting and the white balls are insulating, one can measure percolation electrically; but as the Marseilles group (Clerc *et al.*, 1983), see reading list for Introduction) found out, such experiments are more difficult than the visual inspection recommended here.

For all these lattices, each site is randomly occupied with probability p and empty with probability $(1 - p)$ and clusters are groups of neighbouring occupied sites.

Everything we have defined so far is called 'site percolation'. Its counterpart is called 'bond percolation' and is defined as follows. Imagine each site of the lattice to be occupied, and lines drawn between neighbouring lattice sites. Then each line can be an open bond with probability p, or a closed bond with probability $(1 - p)$. (Simply identify yourself with a water molecule in a coffee percolator; you then can only flow through the open channels, not through the closed ones.) A cluster is then a group of sites connected by open bonds. When measuring the size of a cluster one has to define whether one counts the site content or the bond content. For example, are two sites, which are connected by an open bond with each other and by closed bonds with all other sites, called a cluster of size two (site) or of size one (bond)? Because of this ambiguity this book deals mostly with site percolation, even though bond percolation historically came first.

The percolation threshold p_c is that concentration p at which an infinite network appears in an infinite lattice. For all $p > p_c$ one has a cluster extending from one side of the system to the other, whereas for all $p < p_c$ no such infinite cluster exists. In finite systems as simulated on a computer one does not have in general a sharply defined threshold; any effective threshold values obtained numerically or experimentally need to be extrapolated carefully to infinite system size (called the thermodynamic limit by physicists accustomed to thermodynamics). If one has a mathematically exact (or at least plausible) calculation for p_c, then of course no such extrapolations are needed. Moreover, such exact results can be tested to check the reliability of numerical methods.

In Chapter 4 we will deal in greater detail with computer simulations to determine p_c accurately. 'Series' extrapolations will be explained in the next section. Mathematical methods to calculate this threshold exactly are restricted so far to two dimensions, consistent with the experience in the field of phase transitions that three-dimensional problems in general cannot be solved exactly. The review of Essam *et al.* (1978) (in particular section II D), as well as Kesten (1982), explain how two-dimensional thresholds can be derived mathematically for many simple lattices. But progress in this field is not easy. It took about two decades from the first numerical estimates in 1960 for square bond percolation, over non-rigorous arguments that $p_c = 1/2$ exactly, to a mathematical proof. But today we also know $p_c = 1/2$ for the triangular site, $p_c = 2 \sin(\pi/18)$ for the triangular bond, and $p_c = 1 - 2 \sin(\pi/18)$ for honeycomb bond percolation. For the honeycomb site problem, p_c seems to be smaller than $1/\sqrt{2}$; and for square site percolation, no plausible guess for a possibly exact result is known to me at present. For hypercubic lattices in d dimensions, in general p_c does not seem to be $1/(2d-2)$ exactly since $p_c = 1/(2d-1)$ for very large d. Finally, $p_c = 1/(z-1)$ for Cayley trees (Bethe lattice) where every site has z nearest neighbours. Table 1 summarizes exactly and approximately known percolation thresholds in two and three dimensions. The reader interested in precise values should first consult the more recent literature before relying on the estimates in this table. For progress is being continuously made in computer evaluations of percolation thresholds.

Table 1. Selected percolation thresholds for two- and three-dimensional lattices. 'Site' refers to site percolation and 'bond' to bond percolation. In all cases, only nearest neighbours form clusters, and no correlations are allowed between different sites or bonds. If the result is not exact (see text), the error probably affects only the last decimal.

Lattice	Site	Bond
Honeycomb	0·6962	0·65271
Square	0·59275	0·50000
Triangular	0·50000	0·34729
Diamond	0·428	0·388
Simple cubic	0·3117	0·2492
BCC	0·245	0·1785
FCC	0·198	0·119

In all the examples of the table, clusters are defined as groups of nearest neighbours which are occupied or connected by open bonds. One may also allow next-nearest neighbours to form clusters. Then in Figure 1 not only the squares which have one side in common, but also those touching each other only at a corner, belong to one cluster if they are occupied. Even connections over longer distances have been introduced. The percolation thresholds then go to zero if this connection range goes to infinity. One may even get rid of the lattice completely and look at circles distributed randomly on a piece of paper.

Another important variant, which helps us to go continuously from site to bond percolation, is called site-bond percolation. Then the sites of the bond percolation problem are no longer all occupied; only a fraction x of sites is occupied, the rest are empty. Bonds between neighbouring occupied sites are open with probability p, and we look for clusters of occupied sites connected by open bonds. The bond percolation threshold p_c now decreases from unity, if x equals the site percolation threshold, to the normal bond threshold if $x = 1$.

The occupation of sites has so far been assumed random, which means that each site is occupied or empty independent of its neighbours. Instead one may assume that the sites attract each other and this case has been studied more often than the opposite case of repulsion. If one assumes that each pair of neighbouring occupied sites adds a constant amount to the energy E of the system, then in a real system in thermal equilibrium at the absolute temperature T, the probability of finding the system in a given occupation status is proportional to $\exp(-E/kT)$ where k is Boltzmann's constant. One therefore has to average over the possible configurations of occupied and empty sites even when the total number of occupied sites is fixed. This model is called the lattice gas or Ising magnet and the percolation clusters of neighbouring occupied sites are the Ising clusters. For temperatures T below a certain critical temperature T_c, the occupied sites migrate to one part of the system, the empty sites to another, just as the liquid and vapour phase of a fluid separate below the critical temperature of this fluid. Of course, this phase separation drastically influences the percolative behaviour of the system. In general, one calls all percolation processes 'correlated' if they are not random. The special case just explained is therefore Ising-correlated site percolation.

One may combine the complications of correlation with those of site-bond percolation. For Ising-correlated site-bond percolation, each bond between neighbouring occupied sites in the lattice gas is randomly open with probability p and blocked with probability $(1-p)$. The behaviour of the system now depends on probability p, temperature T and concentration x of occupied sites. This system has a rather complicated combination of phase transitions, a reviewed by Kertész *et al.* in the book *Percolation Processes and Structures* mentioned at the end of Chapter 1. We refer the reader to that article for more details. Other articles there deal, for example, with directed percolation, a problem of particular interest since exact solutions are becoming available, also for directed 'animals'.

In directed bond percolation, an open bond allows a connection only in one direction. Numerous other modifications of percolation have been invented as models for various processes occurring in nature. Many of them were reviewed by Stauffer *et al.* (1982), Chapter 1. The purpose of the present book is to offer an introduction to percolation theory, not a comprehensive review. Thus we will ignore all these complications of correlated percolation, bond percolation, connections with next-nearest neighbours, and other processes, and will work with site percolation on various random lattices in *d* dimensions, except when otherwise stated.

2.2. Exact solution in one dimension

Like so many other problems of theoretical physics, the percolation problem can be solved exactly in one dimension, and some aspects of that solution seem to be valid also for higher dimensions.

Let us study site percolation on an infinitely long linear chain, where 'lattice' sites are placed in fixed distances (Figure 7). Each of these lattice sites is randomly occupied with probability *p*. A cluster is a group of neighbouring occupied sites containing no empty site in between. A single empty site would split the group into two different clusters. In order that the cluster is separated from the other clusters, the site neighbouring the left end of the cluster must be empty; and the same is true for the right end of the cluster. Thus for the central cluster of Figure 7 consisting of five occupied sites, we need these five sites occupied and their two neighbours empty.

Figure 7. Example of clusters in a one-dimensional lattice. The central cluster has five sites, the one to its left is a pair, the one to its right is a cluster of size one, that is an isolated occupied site. The empty sites are not shown.

The probability of each site being occupied is *p*. Since all sites are occupied randomly, the probability of two arbitrary sites being occupied is p^2, for three being occupied is p^3, and for five being occupied is p^5. (This product property of the probabilities is valid only for statistically independent events, as for random percolation, and not for correlated events like the occupation of sites for Ising-correlated percolation.) The probability of one end having an empty neighbour is $(1 - p)$, and again the two ends are statistically independent. Therefore the total probability, that a fixed lattice site is the left end of a five-cluster is $p^5(1 - p)^2$.

How many clusters of size five do we have in our chain, if the total chain length is *L*, with $L \to \infty$, much larger than the cluster length? Every site has the probability $p^5(1 - p)^2$ of being the left hand end of such a cluster, and there are *L* such sites (when we ignore the small number of sites on the end of the whole chain for which the situation is different since there no place is left for five occupied and two empty sites). Thus the total number of five-clusters, apart from effects from the chain ends, is $Lp^5(1-p)^2$. We see that it is practical to talk about the number of clusters per lattice site, which is the total number divided by L and thus $p^5(1-p)^2$. This normalized cluster number thus is independent of the lattice size L and equals the probability that a fixed site is the end of a cluster.

For clusters containing s sites, we define n_s as the number of such s-clusters per lattice site. In our one-dimensional case, the above consideration for five-clusters is easily generalized to

$$n_s = p^s(1-p)^2 \tag{1}$$

This normalized cluster number is crucial for many of our later discussions in two or three dimensions. It equals the probability, in an infinite chain, of an arbitrary site being the left hand end of the cluster. For $p < 1$, the cluster numbers go exponentially to zero if the cluster size s goes to infinity.

The probability that an arbitrary site is part of an s-cluster, and not only its left end, is larger by a factor s; for now that site can be any of the s cluster sites. Thus that probability is $n_s s$. Many authors prefer to work with the probability $n_s s$, instead of with the cluster number n_s. To avoid confusion with the probability p we will not introduce a special symbol for $n_s s$ and will work with the cluster numbers. This cluster number is also the more natural quantity if one counts all clusters in a lattice of a fixed large size by a computer simulation.

Where is the percolation threshold ? For $p = 1$, all sites of the chain are occupied, and the whole chain constitutes one single cluster. For every p smaller than unity, there will be some holes in the chain where a site is not occupied. Thus a chain of length L will have on average $(1-p)L$ empty sites. For L going to infinity at fixed p, this number is also increasing to infinity. Thus there will be at least one empty site somewhere in the chain, and that means that there is no continuous row of occupied sites, i.e. no one-dimensional cluster, connecting the two ends. In other words, there is no percolating cluster for p below unity. Thus the percolation threshold is unity:

$$p_c = 1 \tag{2}$$

Therefore it is not possible to observe the region $p > p_c$ in one dimension. Only one side of the phase transition is accessible since at least this author cannot occupy a site with a probability $p > 1$. Nevertheless, this somewhat unusual phase transition has some similarities with percolation in higher dimensions, and also with certain aggregation processes (Kolb and Herrmann 1984). Thus we will try to squeeze out some more information from this simple result.

The probability that a site belongs to a cluster of size s is $n_s s$, as discussed above. Every occupied site must belong to one cluster since single occupied sites surrounded by empty neighbours are also clusters of size unity. The probability that an arbitrary site belongs to any cluster is therefore equal to the probability p

that it is occupied. Thus

$$\Sigma_s n_s s = p \qquad (p < p_c) \tag{3}$$

The sum runs from $s = 1$ to $s = \infty$. This law can also be checked directly from Equation (1) and the formula for the geometric series:

$$\Sigma_s p^s (1-p)^2 s = (1-p)^2 \Sigma_s p \frac{d(p^s)}{dp}$$

$$= (1-p)^2 p \frac{d(\Sigma_s p^s)}{dp}$$

$$= (1-p)^2 p \frac{d(p/(1-p))}{dp}$$

$$= p$$

For higher dimensions, Equation (3) is also valid except that one has to take into account the sites in the infinite cluster separately, if one does not include them in the sum over all cluster sizes. Therefore Equation (3) is restricted to $p < p_c$; even in one dimension at $p = p_c = 1$ there is only one cluster covering the whole lattice, thus $s = \infty$ and $n_s = 0$, which makes Equation (3) undefined at $p = 1$. (The above trick to calculate a sum by expressing it as a derivative is also useful in other parts of statistical physics.)

How large on average is the cluster we are hitting if we point randomly to a lattice site which is part of a finite cluster? There is a probability $n_s s$ that an arbitrary site (occupied or not) belongs to an s-cluster and a probability $\Sigma_s n_s s$ that it belongs to any finite cluster. Thus $w_s = n_s s / \Sigma_s n_s s$ is the probability that the cluster to which an arbitrary occupied site belongs contains exactly s sites. The average cluster size S which we are measuring in this process of randomly hitting some cluster site is therefore

$$S = \Sigma w_s s$$

$$= \Sigma \frac{n_s s^2}{\Sigma n_s s} \tag{4}$$

Although we will learn later that different types of averages exist, the term mean cluster size is in widespread use for S and will also be used here. (For example, $\Sigma_s n_s s / \Sigma_s n_s$ is the average cluster size if every cluster, and not every site as in Equation (4), is selected with equal probability.) We have defined S here in such a way that Equation (4) is also our definition for higher dimensions provided that the infinite cluster is excluded from the sums.

Let us now calculate this mean cluster size explicitly. The denominator is simply p, as Equation (3) shows. The numerator is

$$(1-p)^2 \Sigma_s s^2 p^s = (1-p)^2 \left(p \frac{d}{dp} \right)^2 \Sigma_s p^s$$

where again the sums go from $s=1$ to infinity, and where the trick from our derivation of Equation (3) is applied twice in order to calculate sums by using suitable derivatives of easier sums. Thus

$$S = \frac{(1+p)}{(1-p)} \qquad (p<p_c) \tag{5}$$

The mean cluster size, diverges if we approach the percolation threshold. We will obtain similar results later in more than one dimension. This divergence is very plausible for if there is an infinite cluster present above the percolation threshold, then slightly below the threshold one already has very large (though finite) clusters. Thus a suitable average over these cluster sizes is also getting very large, if one is only slightly below the threshold.

We may define the correlation function or pair connectivity $g(r)$ as the probability that a site a distance r apart from an occupied site belongs to the same cluster. For $r=0$ that probability $g(0)$ equals unity, of course. For $r=1$ the neighbouring site belongs to the same cluster if and only if it is occupied; this is the case with probability p. For a site at distance r, this site and the $(r-1)$ sites in between this site and the origin at $r=0$ must be occupied without exception, which happens with probability p^r. Thus

$$g(r) = p^r \tag{6}$$

for all p and r. For $p<1$ this correlation function, which is also called a connectivity function, goes to zero exponentially if the distance r goes to infinity.

$$g(r) = \exp \left(\frac{-r}{\xi} \right)$$

where

$$\xi = -\frac{1}{\ln (p)}$$

$$= \frac{1}{(p_c - p)} \tag{7}$$

The last equality in Equation (7) is valid only for p close to $p_c = 1$ and uses the expansion $\ln (1-x) = -x$ for small x. The quantity ξ is called the correlation (or connectivity) length and we see that it also diverges at the threshold. We will see later in higher dimensions that the correlation length is proportional to a typical cluster diameter. This relation is quite obvious here. The length of a cluster with s sites is $(s-1)$, not much different from s if s is large. Thus the average length ξ varies as the average cluster size S:

$$S \propto \xi \qquad (p \to p_c) \qquad (8)$$

Unfortunately we will see later that this relation becomes more complicated in higher dimensions. Rather more generally valid is a relation between the sum over all distances r of the correlation function, and the mean cluster size.

$$\Sigma_r g(r) = S \qquad (9)$$

(The reader who wants to check this and has difficulties should keep in mind that the sum in Equation (9) not only includes $r = 0, 1, 2, \ldots$, i.e. the neighbours to the right, but also the neighbours to the left. They cannot be treated through $r = -1, -2$, however, since r is a non-negative distance. Thus one should calculate the sum over the right neighbours and the centre, $r = 0, 1, 2, \ldots$, multiply it by two to take into account the left part of the lattice, and subtract the contribution unity from the centre, which otherwise would be counted twice.)

We see from this exact solution for one dimension, that certain quantities diverge at the percolation threshold, and that the divergence can be described by simple power laws like $1/(p_c - p)$, at least asymptotically close to p_c. The same seems true in higher dimensions where the problems have not been solved exactly.

The quantities S and ξ have counterparts for other thermal phase transitions. In fluids near their critical point, critical opalescence is observed in light scattering experiments, since the compressibility (analogous to S) and the correlation length ξ diverge there. For magnets, neutron scattering gives similar results, except that now the analogue of our mean cluster size S is the susceptibility. The van der Waals equation for fluids in his thesis of 1873 was the first successful theory to describe aspects of such thermal phase transitions.

One may utilize one-dimensional percolation further by calculating the cluster numbers in finite one-dimensional chains, or by taking into account neighbours more than one lattice distance apart. Then one can check the general concepts of finite-size scaling and universality. But we leave these problems to the experts.

The one-dimensional case now is solved exactly whereas for the d-dimensional case only small clusters will be treated exactly in Section 2.3. There is another case with an exactly known solution, the Bethe lattice, with which we deal in Section 2.4.

2.3. Small clusters and animals in d dimensions

If the one-dimensional solution of Equation (1) is so simple, why can't we apply the same principle to higher dimensions and find the exact solution there? To answer that question, let us look again at the square lattice of Figure 1. First, what is the probability that an arbitrary site is a cluster of size $s = 1$, i.e. an isolated occupied site? For this purpose, the site itself has to be occupied (probability p) while its four neighbour squares have to be empty (probability $(1 - p)$ for each). Again the occupation of these five sites happens independently, and thus the combined probability is the product $n_1 = p(1 - p)^4$. The number of pairs, n_2, can also be calculated easily: Two sites have to be occupied, their six neighbour squares have to be empty, and the pair can be oriented either horizontally or vertically. Thus the average number of pairs per lattice site is $n_2 = 2p^2(1 - p)^6$. Similarly, three sites on a straight line have 8 neighbours, and the average number (per lattice site) or such clusters is $2p^3(1 - p)^8$. Generally, the number of clusters of s sites forming a straight line is $2p^s(1 - p)^{2s + 2}$ on a square lattice, since each such cluster has $(2s + 2)$ empty neighbour squares.

In three dimensions, on the simple cubic lattice, each straight cluster with s sites has $(4s + 2)$ empty neighbours, and three orientations are possible, leading to an average cluster number (per lattice site) of $3p^s(1 - p)^{4s + 2}$.

In a d-dimensional hypercubic lattice, each site has $2d$ neighbours, and for the sites in the interior of an s-cluster forming a straight line, $(2d - 2)$ of these sites have to be empty. Including the two end points, the s-cluster has in this case $2 + (2d - 2)s$ empty neighbours, resulting in a cluster number $dp^3(1 - p)^{(2d - 2)s + 2}$.

This general d-dimensional result includes the above cluster numbers for $d = 2$ and $d = 3$ as well as the one dimensional result ($d = 1$) of Equation (1). Have we thus solved the percolation cluster problem exactly in d dimensions, leaving the evaluations of mean cluster size and correlation length as an exercise analogous to the one-dimensional case?

Unfortunately, our straightforward, exact, simple and complete solution has one slight disadvantage:

IT IS WRONG

The world is not straight. Three sites of a cluster on a square lattice do not necessarily follow a straight line; they can also form a corner, as shown here:

```
        o
    o   x   o
o   x   x   o
    o   o
```

The three occupied cluster sites are marked by an x, the seven empty neighbour sites by an o. Four orientations of this corner are possible; thus the average number (per lattice site) of such corners is $4p^3(1-p)^7$.

Combined with the above result for straight lines we thus get $n_3 = 2p^3(1-p)^6 + 4p^3(1-p)^7$ for the average number of triplets on the square lattice. Figure 8 shows the 19 possible configurations for $s=4$ on the square lattice; it is a nice classroom exercise to find, in a long collaborative effort, all 63 configurations for $s=5$.

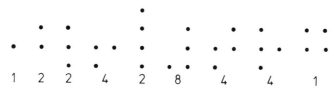

Figure 8. List of all cluster configurations ('lattice animals') on the square lattice up to $s=4$. For each structure, mirror images and rotated configurations are not shown; only the total number of such configurations, including the one shown, is indicated under each structure.

Such an exercise will convince you that the configurations of larger clusters are counted faster and more reliably by a computer. Up to $s=24$ on the square lattice this has been done by Redelmeier who kept a PDP 11/70 computer busy for ten months to count the 10^{13} configurations. A short and complete Fortran program for this purpose was published by Redner. One calls these configurations 'lattice animals' since they have a certain similarity with multicellular living beings which might enter your nightmares if you counted them too long.

But for our purpose it is not sufficient to count these animals; we have to classify them according to the number of empty neighbours each of them has. For example, of the six triplet configurations, two have eight empty neighbours and the remaining four have seven. This difference entered into our above result for the average number of triplets. Generally, the number of empty neighbours of a cluster is called its *perimeter* for which we use the symbol t here (not to be confused with time, of course). (One should not identify the perimeter with a cluster surface since t includes internal holes, like in a Swiss cheese.) Triplets thus have the perimeters $t=7$ or $t=8$ on the square lattice. If the number of lattice animals (cluster configurations) with size s and perimeter t is denoted by g_{st}, then

$$n_s = \Sigma_t g_{st} p^s (1-p)^t \qquad (10)$$

is the average number of s clusters per lattice site. All our above formulae for cluster numbers are special cases of this general formula, which is valid for every lattice. For example, the triplets ($s=3$) on the square lattice have perimeter $t=8$ ($g_{st}=2$ configurations) or $t=7$ ($g_{st}=4$ configurations), and the total cluster number is the sum of these contributions.

The difficulty with Equation (10) is that it involves a sum over all possible perimeters t, and thus each possible configuration has to be found and carefully analysed to find the g_{st}. Tables of such animal numbers, often in the form of so-called perimeter polynomials

$$D_s(q) = \frac{n_s}{p^s}$$

$$= \Sigma_t g_{st} q^t \qquad \text{where } q = (1-p)$$

have been published mainly by the King's College group (see reading list). There seems to be no exact solution for general t and s available at present, and that's why the percolation cluster problem has not yet been solved exactly.

Nevertheless, some asymptotic results are known for very large animals. They are listed here without proof since according to our present knowledge these animals, even if domesticated by exact solutions, do not help us in an exact solution for percolation clusters at the threshold. The perimeter t, averaged over all animals with a given size s, seems to be proportional to s for $s \to \infty$. Thus it is appropriate to classify different animals of the same large size s by the ratio $a = t/s$. If a is smaller than $(1 - p_c)/p_c$ on any lattice in more than one dimension, then

$$g_{st} \sim \left[\frac{(a+1)^{a+1}}{a^a} \right]^s \tag{11}$$

for large animals, apart from factors varying less strongly with s (B. Souillard, unpublished; see the reviews of Stauffer and Essam mentioned after the introduction.) Therefore also the total number $g_s = \Sigma_t g_{st}$ of animals, irrespective of their perimeter, increases exponentially with animal size s, apart from less strongly varying prefactors:

$$g_s \propto s^{-\theta} \text{const}^s \tag{12}$$

Here $\propto$ stands for proportional and is used in this book nearly always in the sense of an asymptotic law, for example for $s \to \infty$, or for $p \to p_c$. In two dimensions, $\theta = 1$ whereas $\theta = 3/2$ in three (Parisi and Sourlas 1981); $\theta = 5/2$ for d above 8, as in the Bethe lattice. Finally, the average radius or diameter of big animals increases as the square root of the animal mass s for large s in three dimensions, in contrast to the impression you may have got in the zoo. (No exact solution for two-dimensional animal radii is known thus far, one of the rare cases where $d = 3$ is better known than $d = 2$.)

Inspection of Equation (10) tells us that averages over percolation clusters of one fixed size correspond, in the limit $p \to 0$, to average over lattice animals. For then the factor $(1 - p)^t$, by which percolation clusters with different perimeters t are distinguished from animals, approaches unity and thus can be omitted. Thus for very small p, the average squared radius of percolation clusters also varies as s

for large s in three dimensions. The numbers of such large percolation clusters are very small, of course, as follows from Equation (12).

$$n_s(p \to 0) \propto s^{-\theta} p^s \ const^s \qquad (13)$$

which goes to zero rapidly for increasing s if $p \ll 1/const$. Thus the percolation clusters, to which our animal limits apply, are very rare. Nevertheless they have been studied numerically by looking at the perimeter polynomials or even by computer simulations, provided one tells the computer to work with one cluster of one fixed size s. One cannot simulate them accurately by just producing clusters randomly with $p = 0.00001$. (Similarly you can study one rare plant seed under a microscope by catching one in the field, putting it into the view of your microscope, and then preventing it from being blown away. If instead you wait until this specimen appears accidentally in your microscope, you may not finish your thesis or project in time.) This somewhat artificial animal limit of $p \to 0$ at fixed size s may be relevant for the study of branched polymers in a very dilute solution.

2.4. Exact solution for the Bethe lattice

Besides the one-dimensional case, another case can also be solved exactly, which in some sense corresponds to infinite dimensionality: the Bethe lattice (or Cayley tree) of Figure 9. The Bethe approximation is a method used to treat magnetism and works exactly on Cayley trees; that is why physicists call these structures 'Bethe lattices'.

What has the Bethe lattice to do with infinite dimensionality d? For $d = 2$, the

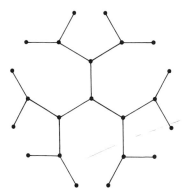

Figure 9. A small Cayley tree or Bethe lattice, where each site except the many surface sites has $z = 3$ neighbours. Percolation theory started with the exact solution on this somewhat artificial structure.

area of a circle with radius r is πr^2 whereas its circumference is $2\pi r$. The surface of a three-dimensional sphere of radius r is $4\pi r^2$ whereas its volume varies as r^3. In d dimensions, the volume of a 'sphere' is proportional to r^d, its surface to r^{d-1}. Thus

$$surface \propto volume^{(1-1/d)} \tag{14}$$

in d dimensions. We see that in the limit $d \to \infty$ the surface becomes proportional to the volume; this is also true if we look at squares, cubes, 'hypercubes' etc.

For the Bethe lattice in Figure 9, one starts with a central point ('origin'), having z bonds, with $z = 3$ in the example of Figure 9. Each bond ends in another site from which again z bonds emanate; one of these z bonds is the connection with the origin, the other $(z-1)$ bonds lead to new sites. This branching process is continued again and again. Thus if we have reached one site in the interior of a Bethe lattice, then we can go on in $(z-1)$ other directions in addition to the direction from which we came. Only at the surface of the lattice, where the branching has stopped, is only one bond connecting the surface site to the interior ('dead end road'). There are no closed loops in this structure, which means that we always reach new sites if we never go back.

We see from our Figure 9 that the number of sites increases exponentially with the distance from the origin, whereas in any d-dimensional structure it would increase with $(distance)^d$. In our example with $z = 3$, the origin is surrounded by a shell of three sites ('first generation'), in the second shell we have six sites, followed by a third generation of twelve sites etc. Thus a sphere of r generations contains $1 + 3(1 + 2 + 4 + \cdots + 2^{(r-1)}) = 3\,2^r - 2$ sites, of which the last generation of $3\,2^{(r-1)}$ sites are surface sites. Thus for large r half of the sites are surface sites, the other half are in the interior of the sphere. With z instead of three neighbours for each site, the fraction of surface sites approaches $(z-2)/(z-1)$, as an analogous calculation shows. Thus the ratio of surface to volume approaches a finite limit. Equation (14) shows that this special behaviour occurs for $1/d = 0$ only, that is for infinite dimensions. And we see also that the Bethe lattice is something very peculiar. When we now talk about percolation in the Bethe lattice, we therefore always have in mind the behaviour in the interior of the Bethe lattice, and not the effects due to the surface which are also important.

Let us now find the percolation threshold in the Bethe lattice. We start at the origin and check if there is a chance of finding an infinite path of occupied neighbours, starting from that origin. If we go on such a path in the outward direction, we find $(z-1)$ new bonds emanating from every new site, apart from the direction from which we came. Each of these $(z-1)$ bonds leads to one new neighbour, which is occupied with probability p. Thus on average we have $(z-1)p$ new occupied neighbours to which we can continue our path. If this number $(z-1)p$ is smaller than unity, the average number of different paths leading to infinity decreases at each generation by this factor <1. Thus even if all z neighbours of an occupied origin happen to be occupied, giving us z different chances to find a way out, and even if z is very large, the probability of finding a continuous path of occupied neighbours goes to zero exponentially with path length, if $p < 1/(z-1)$. Therefore we have derived

$$p_c = \frac{1}{z-1} \tag{15}$$

for the Bethe lattice with z neighbours for every site. (The above argument is also valid if each bond between neighbour sites is open or blocked randomly; thus Equation (15) is valid for both bond percolation and site percolation.)

But even if p is larger than the percolation threshold $1/(z-1)$, the origin does not always have a connection to infinity. For example, if it is occupied and its z neighbours are empty, then it does not belong to the infinite network. We define the percolation probability P as the probability that the origin or any other arbitrarily selected site belongs to the infinite cluster. Clearly this probability is zero for p below the percolation threshold p_c, and we want to calculate it therefore only for $p > 1/(z-1)$. We will see later that this quantity P also makes sense for general lattices, not only for Bethe lattices. (In one dimension our p is never $> p_c$ and thus we did not introduce it there.) To distinguish the two probabilities P (probability of an arbitrary site belonging to the infinite network) and p (probability of an arbitrary site being occupied) we may also call P the strength of the infinite network and p the concentration. But 'strength' here only means the relative amount and should not be confused with the elastic property.

Figure 10 shows the immediate surroundings of the origin and defines what we mean by branch, neighbour, and sub-branch. We want to calculate the strength of the infinite network, that is the probability P that the origin (or any other site) is connected to infinity by occupied sites. We call Q the probability that

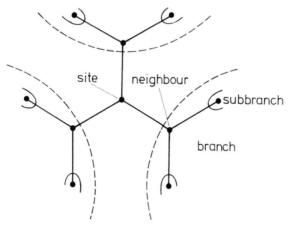

Figure 10. The surroundings of the origin of a large Bethe lattice. This figure defines what we mean by neighbour, branch, and sub-branch in our derivation of the exact solution.

an arbitrary site is not connected to infinity through one fixed branch originating from this site. Taking $z = 3$ for simplicity, as we have done in Figure 10, we now calculate Q from the rule that probabilities for statistically independent events are simply multiplied by each other. The probability that the two sub-branches which start at the neighbour are not both leading to infinity is Q^2. (A sub-branch is connected to infinity with the same probability as a branch since all sites are equivalent in the interior of the Bethe lattice.)

Thus pQ^2 is the probability that this neighbour is occupied but not connected to infinity by any of its two sub-branches. This neighbour is empty with probability $(1 - p)$, in which case even well connected sub-branches don't help it. Thus $Q = 1 - p + pQ^2$ is the probability that this fixed branch does not lead to infinity, either because the connection is already broken at the first neighbour, or because later something is missing in a sub-branch.

This quadratic equation for Q has two solutions $Q = 1$ and $Q = (1 - p)/p$. The probability $(p - P)$ that the origin is occupied but not connected to infinity through any of its three branches is pQ^3. Thus $P = p(1 - Q^3)$ which gives zero for the solution $Q = 1$ (apparently belonging to $p < p_c = 1/2$), and gives

$$\frac{P}{p} = 1 - \left[\frac{1-p}{p}\right]^3 \tag{16}$$

for the other solution, which corresponds to $p > p_c = 1/2$.

Figure 11 displays this result, which goes back to Flory (1941). For it is in polymer chemistry that the first percolation theory was developed by studying bond percolation on this Bethe lattice. (As we saw above, for this special case the difference between bond and site percolation is not very important.) P is then identified with the fraction of atoms which belong to the infinite network. For example, if one prepares a pudding then it is first a fluid (no elasticity, finite viscosity, finite macromolecules, finite clusters, $p < p_c$). After some time it is a jelly with a finite elasticity, and no longer fluid. This process is called gelation and is

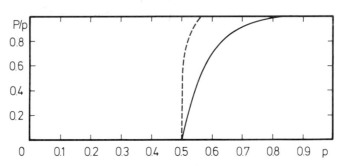

Figure 11. Order parameter P (strength of the infinite network) versus concentration p, in the Bethe lattice with $z = 3$, Equation (16) shown as solid line, and from computer simulations of the triangular lattice (dashed line). In both cases the threshold is at $p_c = 1/2$.

also observed if an egg is boiled for breakfast, if milk becomes cheese, rubber is vulcanized etc. Flory identified this polymerization process with percolation and solved the percolation problem by using the Bethe lattice, which approximates reality by not allowing any closed loops. Indeed, certain Scottish cows have learned since then to follow Equation (16) quite closely when their milk is transformed to cheese. For more gelation details we refer to the review of Stauffer, Coniglio and Adam (1982) listed after the Introduction.

Readers outside France may be less interested in cheese than in magnets and might be acquainted with the concept of the spontaneous magnetization. For other thermal phase transitions the analogue of the spontaneous magnetization is called the order parameter. For temperatures above a certain critical temperature T_c, the order parameter is zero (paramagnetism) whereas for lower temperatures it is non-zero (ferromagnetism). The order parameter for percolation can therefore be identified with the strength P of the infinite network, which is zero for $p < p_c$ and non-zero for $p > p_c$. Thus in analogies with thermal phase transitions one has to be careful about the sign: $p - p_c$ corresponds to $T_c - T$ and not to $T - T_c$. Of course, no analogy is perfect: the order parameter in a ferromagnet (the spontaneous magnetization) can show in at least two different directions, whereas the probability P has a unique value in the 'ferromagnetic region' $(p > p_c)$.

Just as in one dimension, we can also calculate the mean cluster size S for the Bethe lattice. This is the average number of sites of the cluster to which the origin belongs. Again we take $z = 3$ for simplicity. Let T be the mean cluster size for one branch, that is the average number of sites to which the origin is connected and which belongs to one branch. Again, sub-branches have the same mean cluster size T as the branch itself. If the neighbour is empty (probability $(1 - p)$), the cluster size for this branch is zero. If the neighbour is occupied (probability p), it contributes its own mass (unity) to the cluster, and adds the mass T for each of its two sub-branches. Thus

$$T = (1 - p)0 + p(1 + 2T)$$

with the solution $T = p/(1 - 2p)$ for p below the threshold $1/2$. The total cluster size is zero if the origin is empty, and $(1 + 3T)$ if the origin is occupied; therefore the mean size is

$$S = p(1 + 3T)$$

$$= p(1 + p)/(1 - 2p) \tag{17}$$

We have thus derived exact formulae for the mean cluster size S below the percolation threshold, Equation (17), and the strength P of the infinite cluster for concentrations p above the percolation threshold, Equation (16).

If there is an infinite cluster above p_c it is plausible that slightly below p_c the mean cluster size is very large and indeed the denominator in Equation (17) vanishes for $p = 1/2 = p_c$, giving

$$S \propto \frac{1}{(p_c - p)} \qquad (18a)$$

if p approaches p_c from below. On the other hand, if there is no infinite network below p_c it is very possible that slightly above p_c one has only a very weak infinite network, that is P is very small. Indeed, Equation (16) tells us that $P = 0$ at $p = 1/2$, and

$$P \propto (p - p_c) \qquad (18b)$$

if p approaches p_c from above. (To obtain the proportionality factor in Equation (18b) you should make a Taylor expansion of Equation (16) in powers of $p - \frac{1}{2}$; I was too lazy to do that.)

Equation (18) is an example for critical phenomena: Quantities of interest go to zero or infinity by simple power laws. We will discuss similar power laws later when we go to regular d-dimensional lattices instead of the Bethe lattice, only then the power laws are not so simple, with P for example vanishing as $(p - p_c)^{5/36}$ in two dimensions. The power laws are so particularly simple in the Bethe lattice because it is exactly solved (you may also see it the other way round); a simple formula like Equation (16) can hardly give a critical exponent 5/36.

Critical phenomena also occur for thermal phase transitions; the Bethe lattice approximation for percolation theory then is somewhat analogous to the molecular field approximation for magnetism, or the van der Waals equation for fluids. (More precisely, one should use the better Bethe approximation for magnetism which is exact on the Bethe lattice but has the same critical exponents as the molecular field approximation.) In all three cases, rather simple formulae for the order parameter can be derived exactly, leading a simple power law very near to the critical point. For example, the difference between liquid and vapour density, or the spontaneous magnetization, wa: both approaches as $(T_c - T)^{1/2}$. Thus the analogy between thermal critical ph and percolation is not complete since the critical exponent for the o eter is 1/2 for thermal phase transitions and unity for perc these approximations.

In both the thermal phase transitions and percolation th parameter goes to zero continuously if one approaches the critical point. Such phase transitions are called continuous phase transitions or second order phase transitions. If instead the order parameter jumps to zero one has a first order phase transition. Such transitions can occur in more complicated situations but up to now have not played a major role in percolation research. They arise when investigating Ising-correlated site percolation as a function of chemical potential or magnetic field at fixed T below T_c, when looking at the so-called bootstrap percolation, or when studying balloons on a lattice all growing with the same speed.)

The above derivations make clear why it is important to have a Bethe lattice with no loops as in the square lattice, only branches as in a tree. If there had been any connections between the different branches or sub-branches, except for the site or neighbour where the (sub-)branches separate, then we could not have calculated the probabilities for different branches together as products of probabilities for each branch separately. QQQ would no longer have been the probability that none of the three branches is connected to infinity, had the three branches not been statistically independent.

Another reason why Bethe lattices are easier to solve than square lattices becomes clear if one calculates $n_s(p)$, the average number (per site) of clusters containing s sites each. As in one dimension and in contrast to two or three dimensions, the size s of a cluster is uniquely related to its perimeter t, that is to the number of empty neighbours of occupied cluster sites. A single isolated site has three empty neighbours in Figure 9, a pair has four. In the general case of z neighbours per site, the isolated bachelor is surrounded by z perimeter sites whereas the married couple has $(2z - 2)$ empty neighbours. Each child added to t cluster increases t by $(z - 2)$. Thus $t = (z - 2)s + 2$ is the total perimeter of s-ers in the Bethe lattice, just as in the square or simple cubic lattices for t-line clusters. We see that for large s the perimeter is proportional to s, as e case in more realistic lattices (see below). Moreover, the asymptotic uals $(1 - p_c)/p_c$ since this ratio is $(z - 2)$ and since $p_c = 1/(z - 1)$. We will ' this relation is valid at the percolation threshold for all lattices, not ?ethe lattice.

'y our general result given by Equation (10) for the cluster $t = (z - 2)s + 2$:

$$n_s = g_s p^s (1 - p)^{2 + (z - 2)s}$$

(1 s on lu⋅⋅⋅1 . is possible we do not have to sum over t.) Since ag. 'azy, ∴ $z = 3$ for simplicity (similar results hold generally), avoid the ca f the number g_s of different configurations for s-clusters on the Bethe instead look at the ratio (Wingen, unpublished seminar talk)

$$\frac{n_s(p)}{n_s(p_c)} = \left[\frac{(1 - p)}{(1 - p_c)}\right]^2 \left[\left(\frac{p}{p_c}\right)\frac{(1 - p)}{(1 - p_c)}\right]^s$$

$$= \left[\frac{(1 - p)}{(1 - p_c)}\right]^2 [1 - a(p - p_c)^2]^s$$

$$\propto \exp(-cs) \tag{19}$$

Here $a = \frac{1}{4}$

$$c = - \ln \left[1 - a(p - p_c)^2 \right]$$

$$\propto (p - p_c)^2$$

We see that a very simple exponential decay is obtained in Equation (19) for this ratio of cluster numbers. Later we will see that this simplicity is a peculiarity of the Bethe lattice. In two or three dimensions a decay law as $\exp(-cs)$ is valid only for large clusters, and only for $p < p_c$. We call this decay as $\exp(-cs)$ 'animal-like' since for very small p, that is in the animal limit of Section 2.3, the cluster numbers decay as $(p \; const)^s$, Equation (13). The Bethe lattice clusters are always animal-like, whereas in two or three dimensions, only very large percolation clusters below p_c behave like animals.

We now want to find the asymptotic behaviour of the cluster numbers at the threshold, $n_s(p_c)$. We have seen generally, not only for Bethe lattices, in Equation (4) that

$$S \propto \Sigma_s s^2 n_s$$

since the denominator remains finite at the threshold. Thus for $p = p_c$ this ~~~ (also called the second moment of the cluster size distribution) is infinite, wh~~ for any other p it remains finite. If $n_s(p_c)$ decayed exponentially with ~~~ the mean cluster size S would remain finite at $p = p_c$. Thus a power law ~~~ more plausible and defines the Fisher exponent (Fisher droplet mod~~ through

$$n_s(p_c) \propto s^{-\tau}$$

for large s. Again this law is valid rather more generally, not only in the ~~ lattice.

Let us now evaluate S and then calculate τ by comparing the result wit~~ Equation (18). The calculation which follows now uses tricks which occur again and again in the scaling theory of percolation clusters. We assume p to be only slightly smaller than p_c:

$$S \propto \Sigma s^2 n_s$$

$$\propto \Sigma s^{2-\tau} \exp(-cs)$$

$$\propto \int s^{2-\tau} \exp(-cs) \, ds$$

$$= c^{\tau-3} \int z^{2-\tau} \exp(-z) \, dz$$

$$\propto c^{\tau-3}$$

$$\propto (p_c - p)^{2\tau-6} \tag{21}$$

where we used the above result that c vanishes quadratically in $p - p_c$. Equation (18) shows that S must diverge with an exponent -1:

$$2\tau - 6 = -1$$

Thus the Fisher exponent is

$$\tau = 5/2 \tag{22}$$

in the Bethe lattice. We thus can rewrite Equation (19) for large s:

$$n_s(p) \propto s^{-5/2} \exp(-cs) \qquad c \propto (p - p_c)^2 \tag{23}$$

where the first proportionality holds for all p and large s, the second one only for p near p_c.

Of course, one can also derive Equation (23) differently (see for example Essam's review). It is even possible to calculate the cluster numbers at the threshold exactly by calculating the number g_{st} of 'animals' on the Bethe lattice. The resulting expression involving binomial coefficients is however less useful for our later studies of two and three dimensions than the above derivation.

Having derived the Fisher exponent $\tau = 5/2$ from the critical behaviour of the mean cluster size S, we can check if it also gives the correct critical behaviour, Equation (18), of P, the strength of the infinite network. For this purpose we use a general equation, valid for all site percolation problems, not only in the Bethe lattice.

$$P + \Sigma_s n_s s = p \tag{24}$$

where the sum runs over all finite cluster sizes s nd excludes the infinite cluster. This equation simply states that all occupied sites (probability p) either belong to the infinite cluster (probability P) or to one of the finite clusters (probability $\Sigma n_s n$); it generalizes our one-dimensional result, Equation (3), to $p > p_c$ where an infinite network has to be taken into account. (Note again that isolated sites are regarded as clusters of size $s = 1$.) The reader may now check Equation (8) himself, finding help in Equations (28) and (29) using $\sigma = 1/2$ and $\tau = 5/2$.

(Let us mention in passing that besides the Bethe lattice approximation there also exists the effective medium approximation, in particular for percolation properties other than cluster numbers. The critical exponents of this approximation agree neither with those of three-dimensional lattices nor with those of the Bethe lattice, but the behaviour predicted somewhat away from the percolation threshold is quite realistic. For details we refer to Kirkpatrick's review article mentioned after the Introduction.)

In summary, the Bethe lattice solution, Equations (18) and (23) as well as the one-dimensional solution, Equation (1), show that cluster numbers follow rather simple laws, and that exponential decay is quite common. We will utilize these results in the next section to make plausible the scaling law for cluster numbers in general, not just for one and infinite dimensionality.

2.5. Towards a scaling solution for cluster numbers

In this section we try to invent a simple formula which contains the previously discussed one-dimensional and Bethe lattice solutions as special cases. While that formula still is not the final one to be discussed in Section 2.6, it already has many of its properties and in particular already yields the desired scaling laws.

Both Equation (1) for one dimension and Equation (23) for 'infinite dimensionality' (Bethe lattice) are dominated for large cluster sizes s by a simple exponential decay law, $\log (n_s) \propto -s$. Thus we may first postulate for d-dimensional percolation:

$$n_s \propto \exp (-cs)$$

where the factor of proportionality and the parameter c depend on the concentration p. Indeed, if we use the exact result for $s = 1, 2, 3$ in the square lattice, mentioned at the beginning of Section 2.3, we find for $p = 0.1$ the cluster numbers n_s to be 0·06561, 0·01063, and 0·00277 for $s = 1, 2$ and 3, indicating at least a very rapid decay. However, this nice exponential is not consistent with Equation (23), where for the Bethe lattice we also found, in contrast to the one-dimensional case, a power law factor $s^{-\tau}$ by which the exponential is multiplied. Thus instead we postulate more generally:

$$n_s \propto s^{-\tau} \exp (-cs) \tag{25}$$

This law is supposed to be valid for large s only, as was Equation (23). Again, the proportionality factor and the parameter c depend on p whereas we assume the exponent τ to be a constant, though not necessarily equal to 5/2 as in Equation (23). Moreover, near the percolation threshold we no longer require c to vanish as $(p - p_c)^2$ but instead allow a more general power law:

$$c \propto |p - p_c|^{1/\sigma} \qquad (p \rightarrow p_c) \tag{26}$$

Here σ is another free exponent, not necessarily equal to 1/2 as in the Bethe lattice solution given by Equation (23).

Obviously, Equations (25) and (26) are a generalization of our result for the Bethe lattice; does it also contain the one-dimensional result given in Equation (1) as a special case? Using $p_c = 1$ and $p = \exp (\ln p) = \exp (p - 1) = \exp (p - p_c)$ for

$p \rightarrow p_c$, we rewrite Equation (1) as

$$n_s(p) = (p_c - p)^2 \exp\left(-(p_c - p)s\right)$$

Obviously, this result is not a special case of Equation (25) since instead of a power of s we have a power of $p - p_c$ in front of the exponential. We shift the resolution of this discrepancy to the next section by beginning in a more general way. Now we have to choose between a generalization of the one-dimensional result and our generalization Equation (25) of the Bethe lattice solution. We prefer the Bethe lattice as being more realistic than a one-dimensional chain in that it has at least a percolative phase transition: p_c is smaller than unity, in contrast to one dimension, and thus both sides of the threshold can be reached. Thus for the time being we will work with Equation (25).

Equation (25) was assumed to be valid for large s only; perhaps we can get rid of some of the deviations for smaller clusters by investigating the ratio $n_s(p)/n_s(p_c)$, which we call $v_s(p)$ instead of n_s. Then Equation (25) reads

$$v_s \propto \exp\left(-cs\right) \tag{27}$$

where now the exponent τ has cancelled out; it is only implicitly contained in Equation (27) since, from Equation (25),

$$n_s(p_c) \propto s^{-\tau}$$

as in Equation (20). Equation (27) is so simple that it is perhaps worth trusting by the reader; at least even though not entirely correct it allows us to make many calculations quite easily. (You might violate the Official Secrets Act if you now conclude and say loudly that this is what theoretical physicists do. Make calculations if they are easy irrespective of whether the assumptions are correct or wrong.)

First, let us calculate the fraction of sites, P, belonging to the infinite network. A site is either empty or occupied, and if it is occupied it belongs either to a finite cluster (including isolated sites which are treated as clusters of size $s = 1$) or to the infinite network. If we simply set $s = \infty$ in Equation (25) we get zero but that has to be expected. In an infinitely large lattice, which contains at most one infinite network, the number of infinite networks per lattice site is indeed zero. The fraction of lattice sites in the infinite network is calculated by subtracting from the occupied sites those belonging to finite clusters, that is those described by Equation (24). Right at the critical point $p = p_c$ we have $P = 0$ and thus $\Sigma_s n_s s = p_c$. To have this sum converge we need $\tau > 2$. Then we rewrite Equation (24) as

$$P = \Sigma_s [n_s(p_c) - n_s(p)]s$$

$$\propto \Sigma_s s^{1-\tau}[1 - \exp\left(-cs\right)] \tag{28}$$

If p is close to p_c, the factor c in the exponent will be quite small, and only large s values of the order of $1/c$ will give the main contribution to the sum. We may therefore replace the sum by an integral if we are interested only in the leading behaviour near the threshold:

$$P \propto \int s^{1-\tau}[1 - \exp(-cs)]\,ds$$

Integration by parts tell us $\int f'g\,ds = -\int fg'\,ds + (fg)$; here we take

$$f(s) = s^{2-\tau} \qquad g(s) = 1 - \exp(-cs)$$

and get with $z = cs$:

$$P \propto c \int s^{2-\tau} \exp(-cs)\,ds = c^{\tau-2} \int z^{2-\tau} \exp(-z)\,dz$$

(In this definite integral from zero to infinity, the term fg of our integration by parts vanishes since $\tau < 3$.) The integral over z is known as the gamma function $\Gamma(3 - \tau)$ and is available in tabular form. For some applications the reader might need the general rule $\Gamma(x + 1) = x\Gamma(x)$ and $\Gamma(1) = 1$, but here we do not need to know anything about gamma functions since the whole integral is simply a numerical factor (note that τ was assumed to be constant):

$$P \propto c^{\tau-2} \propto (p - p_c)^{(\tau-2)/\sigma} = (p - p_c)^\beta$$

with the critical exponent

$$\beta = (\tau - 2)/\sigma \tag{29}$$

Thus we have found that the much simpler result of the Bethe lattice solution given by Equation (18b) is not valid generally. Instead a critical exponent β is introduced which describes how the strength of the infinite network goes to zero at the percolation threshold. Similarly, for thermal critical phenomena an exponent β describes how the 'order parameter' vanishes. This order parameter is the spontaneous magnetization at the ferromagnet–paramagnet transition, or the difference between liquid density and vapour density at the liquid–gas critical point. In these thermal critical phenomena, β is not in general a simple number like 1 or 1/2, in contrast to the Bethe lattice solution.

Secondly, let us calculate how the mean cluster size S diverges at the threshold. As for Equation (21), we have $S \propto \Sigma s^2 n_s$ since the denominator in Equation (4) remains finite (Equation (24)). The same techniques can be applied, but now we even may avoid the integration by parts:

$$S = \Sigma_s \frac{s^2 n_s}{p_c}$$

$$\propto \int s^2 n_s \, ds$$

$$\propto \int s^{2-\tau} \exp(-cs) \, ds$$

$$\propto c^{3-\tau} \int z^{2-\tau} \exp(-z) \, dz$$

(The careful reader will notice that our calculation neglects the influence of the single infinite cluster. All sums over all cluster sizes from now on are understood to exclude the infinite cluster, if one is present. For one dimension in Equation (4) we did not need that warning since there one cannot have $p > p_c$.) Again the integral over $z = cs$, which equals $\Gamma(3 - \tau)$, is less interesting than the exponent γ for the divergence of S:

$$S \propto c^{\tau - 3}$$

$$\propto |p - p_c|^{(\tau - 3)/\sigma}$$

$$= |p - p_c|^{-\gamma}$$

Thus

$$\gamma = \frac{(3 - \tau)}{\sigma} \tag{30}$$

gives the critical exponent for the mean cluster size S. In order that both β and γ are positive we need $2 < \tau < 3$. For thermal critical phenomena an analogous quantity is the susceptibility of magnets or the compressibility of fluids; both diverge at the critical point with an exponent γ. The numerical value of γ can, of course, be different for different phase transitions. In general the exponent γ is not a simple number like unity for percolation or for thermal critical phenomena. This is in contrast to Equation (18a) for the Bethe lattice and the analogous Curie–Weiss law for the mean-field approximation of the susceptibility.

Does this mean that we have to introduce a new exponent about which we know nothing for every new variable? Obviously this is not the case; for all other quantities derived in this way from the cluster numbers n_s have σ and τ as free parameters in the exponents, but nothing else. Thus if we know these two exponents we know all others. More explicitly, let us calculate the sum

$$M_k = \Sigma_s s^k n_s \tag{31a}$$

which experts also call the k-th moment of the cluster size distribution n_s (if k is an integer). The mean cluster size corresponds to $k = 2$, the strength of the infinite cluster to $k = 1$, and we now allow k to be an arbitrary number $> (\tau - 1)$.

$$M_k \propto \Sigma_s s^{k - \tau} \exp(-cs)$$

$$\propto \int s^{k - \tau} \exp(-cs) \, \mathrm{d}s$$

$$= c^{\tau - 1 - k} \int z^{k - \tau} \exp(-z) \, \mathrm{d}z$$

Thus, apart from a gamma function incorporated into the proportionality factor, we have

$$M_k \propto c^{\tau - 1 - k} \propto |p - p_c|^{(\tau - 1 - k)/\sigma} \tag{31b}$$

Thus the exponent $(k + 1 - \tau)/\sigma$ is again expressed through σ and τ and therefore is not independent of σ and τ. In fact, instead of σ and τ we may also regard β and γ as the fundamental exponents and calculate from them

$$\sigma = 1/(\beta + \gamma), \quad \tau = 2 + \beta/(\beta + \gamma)$$

As the reader can easily check this is the solution of Equations (29) and (30). Thus the critical exponent for the k-th moment is $\beta - (\beta + \gamma)(k - 1)$ according to Equation (31b). Setting $k = 1$ we recover the exponent β for the strength of the infinite network, whereas for $k = 2$ we get $-\gamma$ for the mean cluster size, as it should be.

Some caution is necessary if a sum is not diverging, that is if $k < \tau - 1$. We have this problem already in the evaluation of the first moment; to get the strength P of the infinite network we had to subtract from the sum its value at $p = p_c$, and then replace the sum by an integral. A simple and more general way is to calculate the first derivative (or second, third, ..., derivative, if needed) of the desired sum with respect to c or p. If that derivative diverges, one can replace the sum by an integral, evaluate the result with Equation (31b), and then go back to the original sum. For example, for the first moment $(k = 1)$, Equation (31b) cannot be applied directly since the sum does not diverge. Instead, we calculate

$$-\frac{\mathrm{d}M_1}{\mathrm{d}c} = \Sigma s^2 n_s$$

$$= M_2$$

$$\propto c^{\tau - 3}$$

$$M_1 \propto const + c^{\tau - 2}$$

in agreement with what we derived immediately before Equation (29).

For the zeroth moment $M_0 = \Sigma n_s$ (the total number of clusters), we take the second derivative:

$$\frac{d^2 M_0}{dc^2} = M_2$$

$$\propto c^{\tau - 3}$$

thus

$$M_0 = const_1 + const_2 c + const_3 c^{\tau - 1}$$

The 'singular' or non-analytic part of the total number M_0 of clusters thus varies as

$$M_{0_{sing}} \propto c^{\tau - 1}$$

$$\propto |p - p_c|^{(\tau - 1)/\sigma}$$

$$\propto |p - p_c|^{(2 - \alpha)}$$

with

$$(2 - \alpha) = \frac{(\tau - 1)}{\sigma} = 2\beta + \gamma \tag{32}$$

For thermal phase transitions, the zeroth moment corresponds to the free energy and p is related to the temperature. The second temperature derivative of the free energy is the specific heat and diverges as $|T - T_c|^{-\alpha}$ if the free energy has a non-analytic contribution proportional to $|T - T_c|^{2 - \alpha}$, apart from 'background' terms remaining finite and smooth at the critical temperature.

(Are you doubtful about whether or not you are allowed to evaluate a sum by replacing it directly with an integral? One way to clarify that question is simply to try it. If it does not work you will notice that fact by realizing that the final integral over z (which should be a constant incorporated into the factor of proportionality) does not exist because it diverges at the lower boundary. In that case you should look at suitable derivatives before replacing the sum by an integral. You should not evaluate the sums by Riemann's zeta function unless you are only interested in rough estimates. If you question the accuracy of replacing sums by integrals you might look at the exact example

$$\Sigma_{s=1}^{\infty} s \exp(-cs) = \frac{\exp(-c)}{(1 - \exp(-c))^2}$$

as used already after Equation (3). Approximating the sum by an integral as above we get $1/c^2$. The exact result is more complicated, but for very small c (and this is what we are looking at in a theory of critical exponents), the exact formula can be expanded into $(1 - 5c/12 + \cdots)/c^2$. Thus the leading term is calculated correctly by our approximation.)

Unfortunately, there is something wrong with our assumption. Not only does it fail to include the one-dimensional solution as a special case. It also does not work properly for the strength of the infinite network. Nowhere in our above derivation of the exponent did we actually assume $p > p_c$; thus our formulae would predict an infinite network both above and below the percolation threshold, vanishing only at p_c. Clearly this is wrong, and will be corrected in the following section. To do so we will have to avoid any assumption leading to a maximum of $n_s(p)$ at $p = p_c$ for a fixed large s, when considering the ratio v_s in Equation (27). This maximum must be located below the percolation threshold, for if it is at p_c then $n_s(p_c) - n_s(p)$ is positive both above and below p_c, giving a non-zero strength P of the infinite network (see derivation after Equation (27)) even below p_c.

There is another property of our assumptions in Equations (25) and (26) which should make us suspicious. In Section 2.3 we learned that for a fixed cluster size s the number $n_s(p)$ of such clusters is a finite polynomial in p. Neither $n_s(p)$ nor any of its p-derivatives is allowed to diverge at p_c. But from Equation (26) we find divergences in p-derivatives if $1/\sigma$ is not an integer. For example, if σ is close to 0·4 as in two-dimensional percolation, then $c \propto |p - p_c|^{2 \cdot 5}$, and the third derivative of c and thus of n_s with respect to p diverges roughly as $1/|p - p_c|^{1/2}$. A more reliable assumption therefore has to avoid the expression $z = cs \propto |p - p_c|^{1/\sigma_s}$ as an argument of the exponential. Instead we may try $z \propto (p - p_c)s^\sigma$ and replace Equation (27) by $v_s \propto \exp(-z)$. This assumption is known as the Fisher droplet model and is numerically quite good above the percolation threshold. The whole analysis above can be repeated easily with this droplet model formula; basically only the arguments of the gamma functions are changed, which enter the proportionality factors only. Historically this approach was one of the first scaling theories to thermal critical phenomena and also helped in the application to percolation. As desired, $n_s(p)$ now is perfectly smooth at p_c. But still the Fisher model can hardly be correct below p_c since now $n_s(p)$ goes to infinity, instead of zero, for $s \to \infty$.

(The Fisher droplet model arose from the cluster expansion for real gases and the gas-to-liquid transition. It is similar to earlier theories of nucleation but differs by introducing a surface exponent σ instead of 2/3 and by having a pre-exponential factor varying with $s^{-\tau}$ in the cluster numbers. The scaling theory for the cluster numbers presented here and below can be regarded as a generalization of this Fisher droplet model.)

Nevertheless, our approximation shows the essentials of modern phase transition theory. Everything depends on only two critical exponents. It does not matter whether we call them σ and τ, or β and γ; we only have to keep in mind that from two exponents we can derive the others. For example, from β and γ we can derive the 'specific heat' exponent α via $(2 - \alpha) = (2\beta + \gamma)$. These relationships, known as scaling laws, have been used since the 1960s for thermal phase transitions and in the 1970s were extended to percolation theory. By going through the above formalism the reader will have a better feeling for the more general derivations which follow in the next section.

2.6. Scaling assumption for cluster numbers

If you have read this far through the book it is presumably too late for you to return it and get a refund. Thus now I can tell you the truth: I am unable to offer you the exact solution for the cluster numbers. Instead you are merely offered a further generalization of the above assumption, involving a scaling function so general that everything we discussed so far is contained in it as a special case. No deviations from this scaling assumption have been found (yet) for usual percolation in two and three dimensions.

What have the Fisher model formula, $v_s = \exp\left[-const\,(p - p_c)s^\sigma\right]$ and the simple exponential formula of Equations (25) and (26), $v_s = \exp[-const\,|p - p_c|^{1/\sigma}s]$ in common? In both cases, the function $v_s(p) = n_s(p)/n_s(p_c)$, which depends on the two variables s and $(p - p_c)$, is a function of the combination $|p - p_c|s^\sigma$ only. In the first case that function is an exponential of this combination, in the second it is an exponential of some power of this combination. Thus we may write in both cases:

$$v_s(p) = f(z) \qquad z = (p - p_c)s^\sigma$$

an equation supposed to be valid for p near p_c and large clusters. Inserting the usual law (Equation (20)) at the critical point into this assumption we arrive at our final form

$$n_s(p) = s^{-\tau}f[(p - p_c)s^\sigma] \qquad (p \to p_c, s \to \infty) \qquad (33)$$

The precise form of the scaling function $f = f(z)$ has to be determined by (computer) experiments and other numerical methods and is not predicted by our assumption. While this assumption replaces Equation (25), our previous results (Equations (24), (29–32)) and our definition (Equation (4)) remain valid.

According to the state of the art for several years, this assumption seems to be valid for physical dimensionalities like three-dimensional percolation. In six dimensions, logarithmic correction factors (Essam *et al.*, 1978) slightly modify Equation (33), and above six dimensions the situation may be more complicated. Thus from now on we restrict ourselves to dimensions d between one and six. But before we leave the paradise of very high dimensions we mention that for the Bethe lattice $(d = \infty)$ our Equation (33) is also valid, since it is simply a generalization of Equation (23) via Equation (25).

First let us see whether this new assumption solves the problems we had with the older form (Equation (25)). Does the one-dimensional case of Equation (1) now fit? The one-dimensional result after Equation (26) was incompatible with that earlier assumption. But now we may take $\sigma = 1, z = (p - p_c)s^\sigma$, and rewrite this one-dimensional case as

$$n_s(p) = s^{-2}f(z) = z^2 \exp z \tag{34}$$

valid again for p near unity and s large. Thus we see that now one dimension fits into the picture and corresponds to $\sigma = 1$ and $\tau = 2$. It is somewhat unusual since at $p = p_c$ there are no clusters left. Mathematically this effect results in $f(0) = 0$, but I see no objections why that function cannot be zero at zero argument.

Can we now avoid the appearance of an infinite cluster even below p_c when we apply Equations (24) and (28)? The first part of Equation (28) is still valid, and then we proceed as in the integrations of the preceding section, using $dz/ds = \sigma a/s$ and β from Equation (29):

$$P = \Sigma_s[n_s(p) - n_s(p_c)]s$$

$$= \int s^{1-\tau}[f(z) - f(0)]\, ds$$

$$= |p - p_c|^{(\tau-2)/\sigma} \int |z|^{-1+(2-\tau)/\sigma}[f(z) - f(0)]\, dz/\sigma$$

$$= (\beta + \gamma)|p - p_c|^\beta \int |z|^{-1-\beta}[f(z) - f(0)]\, dz$$

Here the integration over $z = (p - p_c)s^\sigma$ goes from 0 to ∞ for $p > p_c$ and from 0 to $-\infty$ for $p < p_c$. Thus the mystery of the infinite cluster is solved. The scaling function $f(z)$ has to behave, for negative arguments, such that

$$\int |z|^{-1-\beta}[f(z) - f(0)]\, dz = 0$$

or

$$\int |z|^{-\beta}\left[\frac{df}{dz}\right] dz = 0 \tag{35}$$

For positive arguments, that is above p_c, the corresponding integral should not vanish in order to give a non-zero strength of the infinite network. In order to give a vanishing integral (Equation (35)) below the percolation threshold, the function $f(z)$ has to be sometimes larger and sometimes smaller than $f(0)$ and cannot always increase if z increases from $-\infty$ (where f vanishes) to zero. Nature made it simple for us: $f(z)$ has only one maximum, and not many, for usual percolation problems. We call that value of $f(z)$ at this maximum f_{max}, and the negative value of z at this maximum is called z_{max}. Thus

$$f(z_{max}) = f_{max} \qquad f(z) < f_{max} \text{ for } z \neq z_{max} \qquad (36)$$

For a fixed cluster size s, the cluster number n_s thus has a maximum at p_{max} below p_c, with

$$p_{max} = p_c + z_{max} s^{-\sigma} \qquad (37)$$

As a further test, before we go to numerical checks, we want to find out if our assumption (Equation (33)) leads to prohibited divergences in derivatives of cluster numbers with respect to p, as did Equation (25). From what we have assumed so far we cannot exclude that possibility since after all Equation (25) is a special case of our Equation (33) with $\log f \propto -z^{1/\sigma}$. Therefore we now assume in addition that $f(z)$ is an 'analytic' function, which means very (!) roughly that all derivatives of $f(z)$ with respect to z are finite everywhere and in particular at $z = 0$. Since $dz/dp = s^\sigma$ that means also that all derivatives of $n_s(p)$ with respect to p remain finite at $p = p_c$, as they should. Thus the three problems mentioned at the end of the last section seem to be solved.

(Mathematicians might complain that one ought to write $n_s(p) \propto s^\tau f(z)$ with a proportionality factor depending on p. To avoid singularities in the cluster numbers and their derivatives at $p = p_c$ this proportionality factor then should be analytic in p. Near the threshold it therefore differs from its value at the threshold by a term proportional to $p - p_c$. Below p_c this term cannot cancel the $(p - p_c)^\beta$ term derived for the strength of the infinite network before Equation (35), since the exponent β is smaller than unity below six dimensions. For the Bethe lattice, $\beta = 1$, and indeed the Bethe lattice does not obey Equations (35) and (37).)

Finally, let us check if the exponent γ of Equation (30) can now be rederived:

$$S \propto \Sigma s^2 n_s$$

$$\propto \int s^{2-\tau} f(z) \, ds$$

$$= |p - p_c|^{(\tau - 3)/\sigma} \int |z|^{-1 + (3-\tau)/\sigma} f(z) \, dz/\sigma$$

$$\propto |p - p_c|^{-(3-\tau)/\sigma} = |p - p_c|^{-\gamma}$$

Thus Equation (30) has been rederived. The total number of clusters can also be treated in a similar way leading again to Equation (32) and thus to the scaling law

$$(2 - \alpha) = (2\beta + \gamma)$$

as already mentioned above. Of course, these purely theoretical consistency arguments do not yet prove assumption (33). There have been so-called renormalization group arguments in favour of (33) but mainly numerical evidence suggests assumption (33) to be correct.

2.7. Numerical tests

First, let us look at the exact cluster numbers of Section 2.3. For the square lattice we had $n_1 = p(1-p)^4$ for the number of isolated sites and $n_2 = 2p^2(1-p)^4$ for the number of pairs. Do they have a maximum below p_c at fixed cluster size s, as the above argument requires? We find this maximum by setting the p-derivative of n_s equal to zero. For $s = 1$ we thus get

$$(1-p)^4 - 4p(1-p)^3 = 0$$

with the solution $p = 1/5$, whereas for $s = 2$ the same procedure leads to $p = 1/4$. In both cases, the position of the maximum is below the percolation threshold $p_c = 0.5928$; and for the larger cluster the maximum is closer $(1/4)$ to p_c than for the smaller cluster $(1/5)$. This agreement with our theoretical expectation does not yet prove it since our scaling assumption (Equation (33)) is supposed to be valid only for large s. But if one plots $n_s(p)$ using the polynomials as calculated for various lattices by Sykes *et al.* until $s = 10$ to 20, one can determine the position of $p_{max}(s)$ for intermediate s and can show that p_{max} extrapolates for $s \to \infty$ to a value at least very close to p_c, just as equation (36a) requires. In fact, similar methods have even been used to determine p_c quite accurately.

One may also test Equation (33) directly by calculating $v_s(p) = n_s(p)/n_s(p_c)$ from these exact polynomials and plotting the ratio versus $z = (p - p_c)s^\sigma$. Equation (33) then asserts that for different s the results all lie on the same curve $f = f(z)$. (Some people call this effect 'data collapsing'.) Of course, in reality they do not all lie on the curve since even $s = 20$ is rather far from $s = \infty$, and Equation (33) is only assumed to be valid for very large clusters. However, a rough confirmation of Equation (33) has been obtained in this way.

The real strength of the exact polynomials for $n_s(p)$ lies in the determination of critical exponents like β, γ, and σ. For this purpose one is expanding all terms $(1-p)^t$ by the binomial law and orders the result in powers of p. Thus one arrives at a power series

$$M_k = \Sigma_i a_i p^i \tag{38}$$

for the moment M_k one is interested in, as defined in Equation (31a). By looking at the radius of convergence of this series, for example $k = 2$, one finds the percolation threshold p_c. Of course, this determination is inaccurate since only cluster numbers until $s = 10$ to 20 are known. Therefore only the first 10 to 20 terms of the series expansion (Equation (38)) can be calculated. Suitable extrapolation methods have been developed, however, to find the threshold and the critical exponent with great accuracy from a limited number of expansion terms. (See for example the review of Gaunt and Guttmann in vol. 3 of the Domb–Green book series cited after the Introduction.)

One of these extrapolations is called the ratio method. Let us plot the ratios

a_i/a_{i+1} versus $1/i$, and fit a straight line through the data for large i. The intercept A, that is the value of that straight line at $1/i = 0$, then gives p_c. The reason is simple to explain. For large i the terms in the series (38) approach $(p/A)^i$ apart from terms varying less strongly with i. Thus the power series (38) converges for $p/A < 1$ and diverges for $p/A > 1$. Obviously $p = A$ therefore is the critical point: $p_c = A$. With a little bit more effort one can show that the critical exponent for the power series (38) is related to the slope of our plot of ratios versus $1/i$. Unfortunately, for percolation many of the power series have oscillating coefficients a_i for large i, making the ratio analysis invalid.

In these cases, and also others, the so-called Padé approximants still give useful estimates. In this method one approximates the quantity of interest by a ratio of two polynomials

$$\frac{\Sigma_{i=0}^{L} a_i p^i}{\Sigma_{i=0}^{N} b_i p^i}$$

in such a way that an expansion of this ratio in powers of p agrees to all known orders with the exactly known expansion (38). Usually, the best results are obtained when L is close to N and thus about half as large as the total number of known expansion terms. By looking at poles and residues of Padé approximations for derivatives of $\log (M_k)$ with respect to $\log (p)$ one then finds p_c and the critical exponent for M_k. These details as well as the efficient way of determining the expansion coefficients a_i and b_i are beyond the scope of this book.

For percolation, usually these series expansions give the most accurate estimates for the exponents. The determination of the threshold is often less accurate than that by the Monte Carlo method described in Chapter 4. Sometimes the best results are obtained from a Padé approximation which uses a Monte Carlo determination of p_c as input. Of course, when p_c is known exactly, that problem vanishes. Perhaps the greatest triumph of accurate series determinations for percolation was that Domb and Pearce (1976) estimated $\alpha = -0.668 + 0.004$ for triangular site percolation long before $\alpha = -2/3$ was first guessed and then nearly proven by theoretical methods.

None of these series methods is special to percolation. They were all applied earlier to thermal critical phenomena, where a series for the susceptibility of the three-dimensional Ising model gave the first 'non-classical' critical exponent γ from numerical methods. Both ratio and Padé approximations might even be relevant outside phase transition physics, whenever one wants to extrapolate a finite number of data. For example, the birth rate of Canada was reasonably well described by a Padé approximation. The actual numbers from one period are treated as the first series expansion coefficients of some 'generating' function, then a Padé approximation to this function gave all remaining expansion coefficients and thus the birth rates in later years. These birth rates agreed well with the birth rates actually observed.

Now we come to the second method of numerically estimating properties of percolation, the Monte Carlo simulation. For thermal systems, the Metropolis algorithm is an old technique to simulate thermal fluctuations. However for percolation the problem is usually much simpler since the system now is completely random, without memory effects. Thus we may produce a picture like Figure 2 by simply going through the lattice once and occupying each place randomly with probability p. (For thermal critical phenomena one has to go through the lattice thousands of times before some sort of equilibrium is established.) After the production of the configuration one may analyse it by eye or by computer. We shift the latter problem to our Appendix where we describe how one can efficiently count the clusters in such a lattice or check if the system is percolating. Here we concentrate on how to produce a configuration and what to do with the resulting cluster numbers.

Most computers have a built-in random number generator, and even programmable hand calculators often produce random numbers. These random numbers are generated neither at the roulette tables of Monte Carlo, where the name comes from, nor by students who failed in an examination, but by the electrons in the computer circuits. Simple arithmetic or logical operations produce in a completely predictable way a series of numbers which for the outsider, and the simulation, look quite random although they are not. The generation of such pseudo-random numbers is most easily explained on a 32-bit IBM computer, using Fortran language. Take IBM to be a large odd positive integer, called the 'seed' of this particular sequence of random numbers. Then a new odd positive and seemingly random integer can be produced (at least on IBM computers) by

$$\text{IBM} = \text{IBM}*65539$$
$$\text{IF(IBM.LT.0) IBM} = \text{IBM} + 2147483647 + 1$$

The product of IBM and 65539 in general has more than ten decimal digits, that is more than 31 bits and the computer loses the leading digits (bits) and stores only the last 32 bits in the storage place called IBM. The first of these 32 bits is interpreted as the sign of the number, and thus the product of two positive large integers may lead to a negative result. To avoid that accident one adds 2^{31} to IBM if IBM happened to come out negative. The resulting IBM is an odd integer randomly distributed between zero and 2^{31}; it can be used as the input for the next random number to be generated, and so on. To get a real number between zero and unity one has to multiply the integer IBM by $0.465566 \text{ E-}9 = 2^{-31}$. This method of producing random numbers is often available as a subroutine RANDU; but computation time is saved if the statements are written directly into the main program. At most 2^{30} different random numbers can be generated in this way, since there are no more odd integers between zero and 2^{31}. After at most 2^{30} steps, exactly the same sequence of random numbers will be produced again; if you have bad luck, this happens after an even shorter period. In some applications, even with only several thousand lattice sites, problems occur which show that these pseudorandom numbers are not really random. Then a program published by Kirkpatrick and Stoll may help and is nearly as fast. On other computers, where this method may not work, you may use other subroutines, for example the function RANF(i) on CDC computers (where it does not matter at all what the argument of RANF(i) is). In general the production of random numbers is an art and not a science; careful investigations should try different methods to check that the result does not depend on the method used.

As a simple introduction to random numbers and Monte Carlo simulation, let us calculate the number $\pi = 3 \cdot 14159 \ldots$, which is the area within a circle of radius unity. We take a random number which we call x and then another random number called y. Both random numbers are distributed homogeneously between zero and unity. Then we increase a counter, set initially to zero, by one unit if $x^2 + y^2 < 1$; otherwise we leave this counter unchanged. We repeat this computer experiment again and again, increasing the same counter if and only if the resulting $x^2 + y^2$ is smaller than unity. After, say, one million such pairs of random numbers have been used, our counter will be at about $10^6 \pi/4$. The pair (x, y) gives a point within the square of all points with $0 < x < 1$ and $0 < y < 1$. This point lies within the circle of radius unity with probability $\pi/4$. Thus counting the number of points with $x^2 + y^2 < 1$ measures the area of that circle.

How does one occupy a lattice randomly using these pseudo-random numbers? Let us assume we want to have in the array M(20, 20), which represents the 20×20 square lattice of Figure 2, a zero stored for empty places and a one for occupied places, with concentration p. After using the above IBM random number generator we simply state

$$M(I,K) = 0$$
$$IF(IBM.LT.IP) \ M(I,K) = 1$$

where IP is the integer part of $p*2^{31}$. Thus instead of normalizing the random integer IBM by a factor $1/2^{31}$ for each site M(I, K) we only multiply p by 2^{31} once. Then the integer IBM has a probability p of being smaller than IP, and has a probability $(1 - p)$ of being larger. Thus the above statements fill the lattice with the desired probability.

In a finite lattice of, say, one million sites with concentration $p = 0 \cdot 3$ one does not occupy exactly 300 000 sites by this method; it could be several hundred more or less. Different techniques are necessary should one need exactly 300 000 occupied sites. If one merely wants to produce pictures where sites occupied for a lower p are also occupied for larger p, then one can do that easily though not efficiently by restarting with all sites empty for every new p, but always using the same initial seed IBM for the random number generator. Then for different p the same sequence of random numbers is used and if such a random number is smaller than $p = 0 \cdot 3$ (after normalization) it is also smaller than $p = 0 \cdot 4$. Thus the fact that our random numbers are not really random has been used to our advantage, since now a site occupied for small p is automatically also occupied for large p.

Having produced a randomly occupied lattice we may check by eye whether or not it percolates, or we may count clusters. If done manually we will get such bad statistics that a meaningful test of the scaling hypothesis is hardly possible. If we do it on a computer we will get much better results. For $s = 1, 2$, up to 10 or 20 we can plot the results by hand, and then we will get tired and the statistics for n_s will get bad. But up to that size exact cluster numbers are available (see above), and there is no need to produce these clusters by Monte Carlo simulation. Therefore we have to concentrate on larger clusters up to size 1000. Then, as usual in computer work, we must protect ourselves against a flood of data in which we drown before we have analysed it. One way to do that is to combine different neighbouring cluster sizes in one bin. For example, one bin contains all clusters with 8 to 15 sites, the next bin corresponds to s between 16 and 31, then comes the interval 32 to 63, etc, the bin size increasing exponentially. One may plot the result at the geometric mean of the two border sizes s, for example at $s = 45$ for the interval 32 to 63. Then even for very large lattices and very good statistics one does not have too much data to deal with.

Figure 12 shows the cluster numbers for a $95\,000 \times 95\,000$ triangular lattice at the exact $p = p_c = 1/2$, based on one run which took nearly 14 hours on a CDC

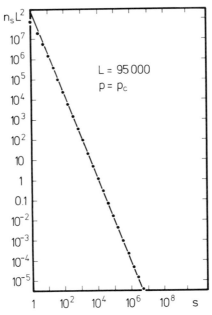

Figure 12. High-quality data on cluster numbers at the percolation threshold, based on one simulation of a 95000×95000 triangular lattice. The slope of the straight line in this log-log plot gives $-\tau$, the Fisher exponent of Equation (20). From Margolina *et al.*, see also Rapaport.

Cyber 76 computer. We see a nice straight line in this log-log plot, corresponding to the Fisher power law, Equation (20). The exponent τ, as given by the negative slope of the straight line, seems to be very close to two, as it should be. Theoretically, $\tau = 2.055$. There are, however, two exceptions. For small s the data points fall below the line since the simple power laws of scaling theory are valid only for large s. And for very large s near one million, the data points seem to be too high, since the boundaries of the lattice cut the infinite cluster into several places, thus enlarging the number of clusters. We see that even a lattice size 95 000 is far from infinity. (Often one can reduce these boundary perturbations by working with periodic boundary conditions. For example, in a 20 × 20 lattice, the right neighbour of the rightmost site $M(i, 20)$ is taken as the leftmost site $M(i, 1)$ in the same row, and the topmost line $(M1, k)$ has the bottom line $M(20, k)$ as top neighbours. This was not done in the example of Figure 12.)

Having tested the validity of scaling theory right at the percolation threshold, we now want to test Equation (33) above and below p_c. Instead of giving high quality data again as reviewed for example in Stauffer's article mentioned after the Introduction, we now take poor data which the reader may easily squeeze out of a medium size computer overnight. Our appendix gives a complete Fortran program and computer output for the simulation of one 500 × 500 lattice at concentrations of 38, 39, 40, ..., 62 per cent. A fast CDC 76 computer needed slightly more than one second for each concentration. We ignore the first four bins in the cluster size distribution, that is the sizes 1, 2–3, 4–7, and 8–15, since for such small clusters scaling is not good. Let us take the fifth bin 16 to 31 as an example. At $p = 0.5$ we found 195 clusters in this size range, at $p = 0.62$ only four. Near $p = 0.39$ is the concentration p_{max} for this size range, with 878 clusters observed, more than at lower or higher concentrations. We take the ratio $n_s(p = 0.39)/n_s(p = 0.50) = 878/195 = 4.5$ and plot this number at $z = (p - p_c)s^\sigma = -0.11 \times (22.6^{0.3956}) = -0.378$, using 22.6 as the (geometric) average cluster size and $\sigma = 36/91$. In a similar way the other data can be processed and lead to Figure 13.

We see in Figure 13 that the different symbols, representing four different size ranges, all follow roughly the same bell-shaped curve. This is exactly what the scaling assumption (Equation (33)) asserted, that all data points follow the same curve $f(z)$. Thus within the very limited accuracy of this test run we have confirmed the scaling theory. Better data, for higher dimensions (Nakanishi and Stanley (1980)), confirm this validity with greater precision. Figure 13 also shows clearly the maximum below the percolation threshold, as opposed to the symmetric scaling function for the Bethe lattice.

If we therefore believe in scaling we may collect the present values of the various critical exponents from the literature. The above scaling laws relate them to each other and to other exponents to be introduced later. Therefore if one exponent is not estimated directly with sufficient accuracy we calculate it from other exponents. In this way, Table 2 gives two- and three-dimensional exponents as well as their 'classical' counterparts for the Bethe lattice. The two-dimensional

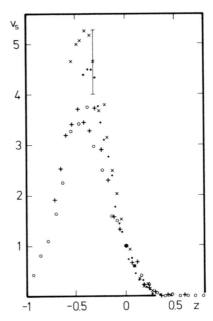

Figure 13. Low-quality test of the scaling assumption (Equation (33)) for the cluster numbers, using the computer output in the Appendix. Different symbols correspond to different size ranges: Dots to 16–31, × to 32–63, + to 64–127 and □ to 128–255. Equation (33) requires all these ratios $v_s(p) = n_s(p)/n_s(p_c)$ to follow the same curve $f(z)$, where $z = (p - p_c)s^\sigma$. Within the strong scattering of the data that rule seems fulfilled.

Table 2. Percolation exponents for two dimensions, three dimensions and in the Bethe lattice, together with the equation number defining the exponent, and the corresponding quantity. Rational numbers give (presumably) exact results whereas those with a decimal point are numerical estimates. The fractal dimension D is denoted by $1/\rho$ in the text.

Exponent	$d=2$	$d=3$	Bethe	Eq.	Quantity
α	$-2/3$	-0.6	-1	32	Total number of clusters
β	$5/36$	0.4	1	29	Strength of inf. network
γ	$43/18$	1.8	1	30	Mean size of finite clusters
ν	$4/3$	0.9	$1/2$	47	Correlation length
σ	$36/91$	0.45	$1/2$	33	Cluster numbers $p = p_c$
τ	$187/91$	2.2	$5/2$	20	Cluster numbers $p = p_c$
$D(p = p_c)$	$91/48$	2.5	4	49 ⎫	
$D(p < p_c)$	1.56	2	4	49 ⎬	Fractal dimension
$D(p > p_c)$	2	3	4	49 ⎭	
$\zeta(p < p_c)$	1	1	1	43 ⎱	Asymptotic exponential
$\zeta(p > p_c)$	$1/2$	$2/3$	1	43 ⎰	decay of cluster numbers
$\theta(p < p_c)$	1	$3/2$	$5/2$	40	Pre-exponential factor
$\theta(p > p_c)$	$5/4$	$-1/9$	$5/2$	42	for asymptotic decay
f_{max}	5.0	1.6	1	36	Max. normalized cl. number
μ	1.3	2.0	3	63	Conductivity

exponents are believed to be exact on the basis of theoretical analogies with thermal phase transitions, which are beyond the scope of this book. The 'striptease' method to be explained in Chapter 4 confirmed them with accuracies of the order of 10^{-4}. The three-dimensional exponents are much less accurate, with no theoretical speculations for their possibly exact values known to me at the time of this writing.

Table 2 does not distinguish between different types of two-dimensional lattice, like square, triangular or honeycomb lattice. All presently available evidence strongly suggests that the critical exponents as well as certain ratios like f_{max} depend only on the dimensionality of the lattice, but not on the lattice structure itself. In other words, if you have seen one two-dimensional lattice you have seen them all for these simple percolation problems. The same is true for d dimensions. Also, bond and site percolation have the same exponents. This simple fact, that critical exponents are independent of the lattice structure, is known in the trade under the complicated name of 'universality' and also holds for thermal phase transitions. For each study of critical exponents it allows us to select that lattice for which our work is easiest.

Our computer output in the appendix also gives the size 'INF' of the cluster and the second moment 'CHI' of the cluster size distribution (excluding the largest cluster). The first is supposed to vanish at the threshold with an exponent β whereas the second should diverge with the same exponent γ whether p_c is approached from below or above. We plot data points in Figure 14; they do not seem to follow a straight line. Of course, for such a 'small' lattice one should not

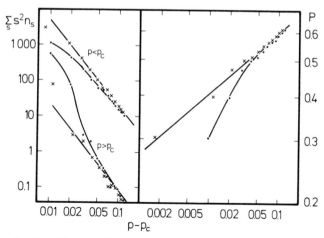

Figure 14. Log-log plot of the strength P of the infinite cluster (top) and of the second moment M_z of the cluster size distribution (bottom) for the triangular lattice, using the low-quality data of the computer output in the Appendix (500×500 triangular lattice). For $p_c = 1/2$, the exact value, the data are difficult to analyse (dots). If p_c is shifted to 0·5083 to take into account some finite size effects (crosses), we get reasonable straight lines with slopes close to the desired exponents.

expect perfect agreement with theory since the boundaries cut smaller pieces off the largest cluster etc. One may repair part of these systematic errors due to the finite system size by working with an effective percolation threshold which differs from the true $p_c = 1/2$ but approaches the true threshold if the system size goes to infinity. Thus we shift a trial value for the effective threshold until the exponent γ is the same above and below the threshold. Using all our data from $p = 0.4$ to $p = 0.6$ we find (via log-log paper, hand calculator, or minicomputer) that for an effective threshold at 0.50827 the two exponents agree and have the value $\gamma = 2.39$, in excellent agreement with the desired $43/18$. Figure 14 shows these data too. By measuring the distance between the two parallel lines on the log-log plot for this shifted threshold value one finds that the second moment of the cluster size distribution is about 200 times larger below the threshold than at an equal distance above the (shifted) threshold, in agreement with determinations from larger lattices. Using this same effective p_c and all data for the largest cluster from $p = 0.51$ to $p = 0.6$, we find the exponent β to be about 0.17, in reasonable agreement with the desired $5/36$. Thus by suitable analysis one can squeeze out good results even from bad data. Also we see here that the usual formulae to estimate error bars for straight-line fits are unreliable, since they ignore the systematic errors due to finite system size (somehow corrected here), finite $p - p_c$ etc.

Besides the Monte Carlo method described here the reader can find other methods to produce clusters, as described by Leath (1976) and by Wilkinson and Barsony (1984).

2.8. Cluster numbers away from p_c

Not everything in life is connected with critical phenomena near p_c. There are also interesting effects in the cluster numbers far away from the percolation threshold. It may be possible that some of their aspects are then approximated reasonably well by the Bethe lattice solution, or by effective medium theories. We concentrate here on (presumably) exact results for large clusters. In a way similar to a good politician, nature provides us with a compromise here: Sometimes it agrees with our Bethe lattice solution and sometimes it disagrees.

First let us argue why for concentrations below the percolation threshold the cluster numbers decay exponentially with cluster size s, that is:

$$\log n_s \propto -s \qquad (s \rightarrow \infty,\ p < p_c) \tag{39}$$

We had seen in Equation (12) that the total number g_s of cluster configurations ('lattice animals') varies exponentially with s, that is

$$\log g_s \propto +s \qquad (s \rightarrow \infty)$$

apart from the less important contribution from the pre-exponential factor. Equation (13) has already explained why an analogous result should be valid for the cluster numbers, for small p, since then $n_s(p) = g_s p^s$. From this result we immediately get Equation (39), as well as less important contributions, logarithmic in s, from the pre-exponential factor. Kunz and Souillard (1978) as well as Schwartz (1978) have shown that this exponential decay is valid for all p below some characteristic value p' which in turn is smaller than p_c. Numerical tests, as performed for example with the exact cluster numbers of section 2.3, already support Equation (39) for medium s.

However, there is a general belief, supported by renormalization group arguments to be discussed later, that Equation (39) is valid for all p below p_c, in other words, $p' = p_c$. Thus we may write this unproven result as

$$n_s(p < p_c) \propto s^{-\theta} \, const^s \qquad (s \to \infty) \tag{40}$$

where we also use an exponent θ, see Equations (12) and (13), for the pre-exponential factor. This exponent is thought to be the same for all p below p_c. Since Equation (40) is supposed to be valid for all p below the threshold it is also valid for small p; therefore everywhere θ equals its values 1 and 3/2 for two and three dimensional animals, respectively, as mentioned after Equation (12).

Nevertheless, it should not be forgotten that this extension of Equations (39) and (40) to all p below p_c is not rigorously proven. There may be examples of non-random percolation where a more complicated behaviour occurs as a function of p. For the simple cases of random site or bond percolation in two and three dimensions no such indications against Equation (40) exist at present.

The behaviour above the percolation threshold is more interesting in that the dimensionality d enters the asymptotic decay law:

$$\log n_s(p > p_c) \propto -s^{1 - 1/d} \tag{41}$$

or

$$n_s(p > p_c) \propto s^{-\theta'} \exp(-Cs^{1 - 1/d}) \tag{42}$$

Again this law is supposed to be valid for large s only. Equation (41) was proven for all p above some p'' which is larger than p_c, and Equations (41) and (42) are thought to be valid for all p above the threshold for simple percolation problems.

The term $s^{2/3}$ (for $d = 3$ dimensions; the exponent is $(1 - 1/d)$ in d dimensions) in the exponential of Equation (42) suggests a behaviour dominated by a surface since the surface area varies as $(\text{volume})^{(1 - 1/d)}$ (see Equation (14)). Indeed, following Kunz and Souillard (1978), we can make this surface term plausible by the following argument. Let us look at the infinite network, to which above p_c a finite fraction of all lattice sites belongs. How much effort is needed to cut out from this network a finite, roughly spherical, cluster of radius r (in units of the lattice spacing)? The geometrical surface of this sphere is $4\pi r^2$. To transform the interior of this sphere into a finite cluster we have to cut all its connections with the outside infinite network. This can be done by making all sites on the surface of the sphere empty, that is less than about $4\pi r^2$ sites. The probability for a fixed configuration of s occupied and t empty sites generally is $p^s(1-p)^t$, as Equation (10) told us. To cut a finite sphere out of the infinite network, the change in both s and t is proportional to the surface area $4\pi r^2$, and the probability that such a cut occurs randomly varies therefore as $\exp(-const\ r^2)$. The number of large finite clusters is at least as large as the number of spherical clusters cut out randomly in this fashion. Therefore n_s cannot be smaller than $\exp(-const\ r^2)$ in three dimensions, provided one has an infinite network present. In d dimensions we get analogously

$$n_s(p>p_c)>\exp(-const\ r^{d-1})\propto\exp(-const\ s^{(1-1/d)})$$

since the number s of network sites in a (hyper)sphere of radius r varies as r^d. This result means that the cluster numbers cannot decay with a simple exponential as in Equation (39) since that decay would be too fast. For large s, s is always larger than any term proportional to $s^{(1-1/d)}$.

This simple argument explains why Equation (39) below the threshold is replaced by Equation (41) above the threshold. From another inequality one can show that $|\log n_s|$ is indeed proportional to $s^{(1-1/d)}$ and not only smaller. Again this latter result is not rigorously proven for all p above p_c but widely believed to be valid in that whole range for simple percolation. Numerical data, for example from the exact cluster numbers at intermediate s, confirm that Equation (41) is a good approximation already for s near 10. (The exact exponent θ' in Equation (42) is 5/4 in two and $-1/9$ in three dimensions according to Lubensky and McKane (1981).) In the Bethe lattice we have no such difference between Equation (39) and Equation (41) but we also see why. The Bethe lattice corresponds to infinite d, and then (and only then) $(1-1/d)$ and 1 are identical. Unfortunately, these exactly known exponents like $(1-1/d)$ have not led to an exact solution for the three-dimensional percolation exponents near the threshold.

There is no contradiction between Equations (39)–(42), presumably exact away from p_c, and the scaling assumption (33), presumably true near p_c. For p close to p_c but s so large that $|z|=|p-p_c|s^\sigma$ is much larger than unity, both Equations (39)–(42) and Equation (33) are expected to be valid. Therefore the scaling function $f(z)$ in Equation (33) must behave for large $|z|$ in such a way that Equation (40) is fulfilled below and Equation (42) is fulfilled above the threshold. For $p>p_c$, for example, we need

$$f(z)\propto z^{(\tau-\theta')/\sigma}\exp(-const'\ z^{(1-1/d)/\sigma})$$

to achieve that aim, with a simpler law below p_c. Monte Carlo data is in reasonable agreement with this seemingly complicated formula (see for example Stauffer's review mentioned after the Introduction.)

We may summarize the main result of this section with the help of an exponent ζ defined through

$$\log n_s \propto -s^{\zeta} \qquad (s \to \infty, p \text{ fixed}) \qquad (43a)$$

Then

$$\zeta(p < p_c) = 1; \qquad \zeta(p > p_c) = 1 - 1/d \qquad (43b)$$

In the next chapter we will investigate whether we also see this difference between above and below p_c (where a surface term only appears above p_c), in the structure of clusters. What this chapter tried to do is, in a somewhat unhistorical way, to explain why percolation cluster numbers have to behave the way they do. Then we resorted to numerical tests to check that we are right. Historically it was much more a case of the opposite: Computers told us that we were wrong until our brains hit the right solution.

Further reading

The King's College group (M.F. Sykes, D.S. Gaunt, M. Glen, H. Ruskin) published their full 'perimeter polynomials', Equation (10), in *J. Phys. A*, **9**, 87, 1705, 1899, (1976), **11**, 1369, (1978) and **14**, 287, (1981). A simple table to read off the resulting cluster numbers for square and simple cubic lattices was published by A. Flammang, *Z. Physik B*, **28**, 47, (1977).

The Metropolis method of Monte Carlo simulation is reviewed by K. Binder (ed.), *Monte Carlo Methods in Statistical Physics*, (Heidelberg: Springer Verlag, 1979) and *Applications of the Monte Carlo Method in Statistical Physics*, (Heidelberg: Springer Verlag, 1984).

Other articles cited in the text are:
Domb, C. and Pearce, C.J., *J. Phys. A*, **9**, L 137, (1976).
Essam, J.W., Gaunt, D.S. and Guttmann, A.J., *J. Phys. A*, **11**, 1983, (1978).
Fisher, M.E., *Physics*, **3**, 255, (1967).
Flory, P.J., *J. Am. Chem. Soc.*, **63**, 3091, (1941).
Kirkpatrick, S. and Stoll, E.P., *J. Computational Phys.*, **40**, 517, (1981). But see Marsaglia, G., Random number generation, in *Encyclopedia of Computer Science*.
Kolb, M. and Herrmann, H.J., *J. Phys. A*, **18**, L (1985).
Kunz, H. and Souillard, B., *J. Statist. Phys.*, **19**, 77, (1978).
Leath, P.L., *Phys. Rev. B*, **14**, 5064, (1976).
Lubensky, T.C. and A.J. McKane, A.J., *J. Phys. A*, **14**, L 157, (1981).
Margolina, A., Djordjevic, Z., Stanley, H.E. and Stauffer, D., *Phys. Rev. B*, **28**, 1652, (1983).
Margolina, A., Nakanishi, H., Stauffer, D. and Stanley, H.E., *J. Phys. A*, **17**, 1683, (1984).
Nakanishi, H. and Stanley, H.E., *Phys. Rev. B*, **22**, 2466, (1980).
Parisi, G. and Sourlas, N., *Phys. Rev. Lett.*, **46**, 871, (1981).

Redelmeier, D.H., *Discrete Math.*, **36**, 191, (1981) as analysed by Guttmann, A.J., *J. Phys. A*, **15**, 1987, (1982).

Redner, S., *J. Statist. Phys.*, **29**, 309, (1982). See also Demmer, E.S. and Diemer, K., *J. Undergrad. Res. in Physics*, (1985) and Sykes, M.F. and Brak, R. (to be published).

Rapaport, D.C., *J. Phys. A*, **18**, L 175, (1985).

Schwartz, M., *Phys. Rev. B*, **18**, 2364, (1978).

Wilkinson, D. and Barsony, M., *J. Phys. A*, **17**, L 129, (1984).

CHAPTER 3

cluster structure

3.1. Is the cluster perimeter a real perimeter?

In Section 2.3 we introduced the 'perimeter' t of a cluster, which is the number of empty sites neighbouring an occupied cluster site. We may call the size s of a cluster, the number of occupied sites, the mass of this cluster; then t is one of the quantities which define the structure of this mass. The word perimeter suggests that it is some sort of surface, similar to the perimeter of a circle which is $2\pi \times$ radius and thus proportional to the square root of the 'mass' (area) πr^2 of the circle. Thus one might expect, at first sight, that t is also proportional to $s^{(1-1/d)}$ in d dimensions, analogous to Equation (14). The aim of this section is to show that this is not so.

We only have to look at Figure 2 to see that the infinite cluster for concentrations p above the percolation threshold p_c has some holes in its interior. Each of these holes gives a contribution to the perimeter. If we have one hole for, say, every thirty sites we have a perimeter proportional to the number of sites in the infinite network. For a very large but finite cluster one may expect the same behaviour as for the infinite network and thus also a perimeter proportional to the number of sites in the cluster. Thus

$$t \propto s \qquad (s \rightarrow \infty)$$

seems plausible according to these arguments. If correct, this quantity t is not a quantity which may be identified directly with a cluster surface.

Do you want to have a proof? Leath (as cited in the previous chapter) has given you one. First we have to define the average perimeter t_s of a cluster containing s sites, for Equation (10) tells us that different clusters with the same mass s have different perimeters. We take:

$$t_s = \Sigma_t \frac{t n_{st}}{n_s}$$

where

$$n_{st} = g_{st} p^s (1-p)^t$$

is the average number of s-clusters having t perimeter sites each, as is obvious from Equation (10); of course, $\Sigma_t n_{st}$ gives n_s. If we differentiate the quantity n_{st} with respect to p we get

$$\frac{dn_s}{dt} = \Sigma_t g_{st} [sp^{s-1}(1-p)^t - p^s t(1-p)^{t-1}]$$

which leads to

$$t_s = s \frac{(1-p)}{p} - (1-p) \, d \frac{\ln (n_s)}{dp} \tag{44a}$$

Now we insert what we learned in the last section, that $\ln (n_s) = -Cs^\zeta$ apart from terms varying less strongly with s, with a p-dependent factor C. Thus also $d \ln (n_s)/dp$ varies as s^ζ, and

$$t_s = s \frac{(1-p)}{p} - \text{const } s^\zeta \qquad (s \to \infty) \tag{44b}$$

We see from Equation (44b) that for sufficiently large clusters the perimeter t_s is always proportional to the mass s. Thus the perimeter is not a surface in the usual sense. Even deep in the interior of the cluster one has perimeter sites, just as holes in a Swiss cheese (or water in a Norwegian fjord) prevent the solid cheese or earth from filling the space completely. Only the second term in Equation (44b) may correspond to a usual surface contribution, since for $p > p_c$ we have $\zeta = (1 - 1/d)$ from Equation (43), giving a perimeter contribution proportional to the usual surface.

You may think that the perimeter does give a real surface if one restricts it to the external perimeter, that is to those sites which are connected by a chain of empty sites to the space far away from the cluster. Indeed, in the Swiss cheese the interior holes have no connection to the outside air and thus do not correspond to the external perimeter. However, for the simple cubic lattice at least, we can see easily that even this external perimeter varies as the volume s, and not as a surface $\propto s^{2/3}$. Let us take p between 0·4 and 0·6. Thus p lies between $p_c = 0·312$ and $(1 - p_c)$. Now not only the occupied sites (concentration p) percolate through the lattice but also the empty sites (concentration $(1 - p)$). Nearly every occupied site is part of the infinite network of occupied sites, and nearly every empty site belongs to the infinite network of connected empty sites. Every large cluster of occupied sites is penetrated by a web of empty sites connected with the outside. Thus even the external perimeter will be proportional to s, and not to $s^{2/3}$. (In two dimensions this argument is not valid, and the external perimeter may play a more important role for percolation.) In summary, neither the perimeter nor the external perimeter measure the surface directly in d dimensions, contrary to what the name 'perimeter' suggests; different definitions are needed to study cluster surfaces in the usual sense (see for examples the papers by Franke (1982)).

3.2. Cluster radius and fractal dimension

While we have seen that 'surfaces' are difficult to define, the 'radius' of a complicated object is much easier to study. Polymer scientists have always had to deal with objects more complicated than a straight line, a square or a sphere. They usually define a 'radius of gyration' R_s for a complicated polymer through

$$R_s^2 = \Sigma_{i=1}^s \frac{|r_i - r_0|^2}{s} \qquad (45a)$$

where

$$r_0 = \Sigma_{i=1}^s \frac{r_i}{s} \qquad (45b)$$

is the position of the centre of mass of the polymer, and r_i the position of the i-th atom in the polymer. We now use the same definition for our percolation problem, replacing polymer by cluster and atom by occupied site. If we average over all clusters having a given size s, the average of the squared radii is denoted as R_s^2. If we turn a two-dimensional cluster around an axis through its centre of mass and perpendicular to the cluster, then the kinetic energy and angular momentum of this rotation is the same as if all sites were on a ring of radius R centred about the axis. Therefore such radii are called 'gyration' radii. We may also relate R_s to the average distance between two cluster sites:

$$2R_s^2 = \Sigma_{ij} \frac{|r_i - r_j|^2}{s^2} \qquad (45c)$$

as one can derive easily after putting the origin of the co-ordinates into the cluster centre-of-mass: $r_0 = 0$.

The correlation function $g(r)$ is the probability that a site at distance r from an occupied site in a finite cluster is also occupied and belongs to the same cluster (see end of Section 2.2). The average number of sites to which an occupied site at the origin is connected is therefore $\Sigma g(r)$, the sum running over all lattice sites r. On the other hand, this average number equals $\Sigma_s s^2 n_s/p$, since $n_s s/p$ is the probability that an occupied site belongs to an s-cluster, that is to a cluster containing s mutually connected sites. Thus

$$pS = \Sigma_s s^2 n_s = p\Sigma_r g(r) \qquad (p < p_c) \qquad (46)$$

the second moment of the cluster size distribution equals the sum over the correlation function (apart from an uninteresting factor p). Above p_c this relation is also valid if the contribution from the infinite cluster is subtracted. Such a relation between mean cluster size (or second moment) and correlation function has already been mentioned in Equation (9). It is also common to systems with thermal fluctuations. For ferromagnets, one may define a correlation function through the probability that two magnetic moments at distance r are parallel, minus the probability that they are antiparallel; then the sum over this correlation function is related to the susceptibility. Near the Curie point this susceptibility diverges with the critical exponent γ just as our second moment or mean cluster size does near the percolation threshold. For percolation, the word 'connectivity' function is also used for our $g(r)$.

We define the correlation or connectivity length ξ as some average distance of two sites belonging to the same cluster:

$$\xi^2 = \frac{\Sigma_r r^2 g(r)}{\Sigma_r g(r)} \tag{47a}$$

Since for a given cluster, $2R_s^2$ is the average squared distance between two cluster sites, since a site belongs with probability $n_s s$ to an s-cluster, and since it is then connected to s sites, the corresponding average over $2R_s^2$ gives the squared correlation length:

$$\xi^2 = \frac{2\Sigma_s R_s^2 s^2 n_s}{\Sigma s^2 n_s} \tag{47b}$$

Thus apart from numerical factors, the correlation length is the radius of those clusters which give the main contribution to the second moment of the cluster size distribution near the percolation threshold. Also for thermal fluctions one defines a correlation length analogous to Equation (47a); this correlation length diverges at the critical point with an exponent v. Thus for percolation we also define an exponent v through

$$\xi \propto |p - p_c|^{-v} \tag{47c}$$

For two-dimensional percolation, plausible but not rigorous arguments give $v = 4/3$, in excellent agreement with numerical results. In three dimensions, v is somewhat smaller than 0·9, whereas for Bethe lattices one has $v = 1/2$, analogous to numerous mean-field theories for thermal phase transitions.

We have seen in Chapter 2 that many quantities diverge at the percolation threshold. Most of these quantities involve sums over all cluster sizes s; their main contribution comes from s of the order of $|p - p_c|^{-1/\sigma}$ (Equation (33)). Now we see that the correlation length, which is also one of these quantities (Equation (47b)), is simply the radius of those clusters which contribute mainly to the divergences. This effect is the foundation of scaling theory. There is one and only one length ξ dominating the critical behaviour. In contrast to politics, this dictatorial principle works quite successfully for all sorts of critical phenomena in two and three dimensions, not only for percolation.

However, the dictatorship principle does not always work even in critical phenomena. If a sum over all cluster sizes s is not diverging at the critical point, then the main contribution to this sum does not come from clusters with radii of the order of the correlation length. Instead the main contribution comes from small s near, say, 10. For example, the sum $\Sigma n_s s$ equals p as long as p is not larger than p_c. Thus it does not diverge at the critical point. At $p = p_c$ the dictatorship principle asserts that the main contribution comes from $s = \infty$; but for the sum over $s^{1-\tau}$ the main contribution comes from rather small s. Then one has to be very careful in the evaluation of critical exponents. Let us take as an example three different definitions of an average squared radius:

$$\frac{\Sigma_s R_s^2 n_s s^2}{\Sigma_s n_s s^2}$$

$$\frac{\Sigma_s R_s^2 n_s s}{\Sigma_s n_s s}$$

$$\Sigma_s \frac{R_s^2 n_s}{\Sigma_s n_s}$$

Here the first expression is our definition of the squared correlation length and divergence with the exponent $2v$. In the second expression the denominator remains finite at the threshold whereas the numerator diverges with the exponent $(2v - \beta)$ (Equation (31b), see also Equation (65) later). In the third expression, in d dimensions, for $d > 2$ both the numerator and the denominator remain finite, and the exponent is 0. Thus depending on the type of averaging the particular situation requires, the critical exponent varies appreciably. In short, do not rely on dictators. (Polymer scientists call the first expression a z-average, the second a weight average, and the third a number average over the squared cluster radius.)

Now we want to find out how the radius R_s varies with s at the percolation threshold. This question leads to the fashionable concept of fractal dimensions. You may have noticed with sadness that a small bottle of scotch, half as high as the customary whisky bottle, does not contain half as much of the precious fluid but only one eighth. For not only the height is reduced by a factor 2 but also the width and the depth, with the volume being the product of these three lengths. In other words, the bottle is a three-dimensional object. For two dimensions, a piece of paper which has a length and a width both twice as large as that of another piece weighs four times as much. Only a one-dimensional object, like a long wire, is simple. If the wire is cut in the middle, each half weighs half as much as the whole wire. In summary, for d dimensions the mass s and the length R_s are related by $s \propto R_s^d$ if one compares geometrically similar objects like bottles, squares or wires. More generally, Mandelbrot defines an exponent D roughly through

$$mass \propto length^D \tag{48}$$

and denotes objects as 'fractals' if they obey Equation (48) with a D different from the 'Euclidean' dimension d of the space in which they are imbedded. We denote $1/D$ by ρ for percolation clusters since we want to use the symbol D for the diffusivity in Chapter 5.

(More precisely, fractals have to be self-similar, as discussed for percolation for example by Kapitulnik et al. (1948). Also, usually one is interested in the asymptotic behaviour for very large lengths, and Equation (48) may be invalid for small objects. Then these objects are fractals only asymptotically. An infinite object like the percolating network is called a fractal if large but finite subsections of it are fractals in the above sense. For our purposes it is more practical to think of fractals as sets of objects of different sizes obeying Equation (48) at least asymptotically, and not to restrict the fractal property to one infinitely large object only.)

The first fractal in this book appeared at the beginning of Section 1.3, for diffusion on an ordered or disordered lattice. We may identify the path of the diffusing particle with the shape of a polymer chain, then the time of the diffusion process is proportional to the mass of the polymer. Thus the mass (time) varies as the square of the length R on the ordered lattice, meaning that the fractal dimension of this normal random walk is $1/\rho = 2$ in both two and three dimensions. For disordered lattices we mentioned in Section 1.3 that a different power law is valid at the percolation threshold, leading to a $1/\rho$ larger than the lattice dimension d. Most fractals, however, have a fractal dimension $1/\rho$ smaller than d.

Are the finite clusters of percolation theory fractals in the sense of Equation (48)? Let us assume

$$R_s \propto s^\rho \qquad (p = p_c, \ s \to \infty) \tag{49}$$

and relate ρ to ν. The denominator of Equation (47b) is the k-th moment of the cluster size distribution with $k = 2$ (Equation (31a)), and diverges with the exponent $\gamma = (3 - \tau)/\sigma$ (Equation (31b)). If near p_c the radius R_s varies as s^ρ, the numerator of Equation (47b) is a moment with $k = 2 + 2\rho$ and thus diverges with the exponent $(3 - \tau + 2\rho)/\sigma$ according to the same Equation (31). Thus their ratio diverges with the exponent $2\rho/\sigma$. This exponent should equal 2ν according to Equation (48). Thus

$$\rho = \sigma\nu \tag{50}$$

This fractal dimension $1/\rho$ is $91/48 = 1\cdot896$ in two and about $2\cdot5$ in three dimensions, and 4 for Bethe lattices. Thus the finite clusters at the percolation threshold are fractals in the sense that their fractal dimension $1/\rho$ is smaller than their lattice dimension d.

For Bethe lattices, Zimm and Stockmayer (1949) showed that $\rho = 1/4$ is the same for all p, not only at p_c. Can we expect also for three-dimensional percolation that ρ is the same above, at and below the threshold? We cannot. Imagine we have p very close to unity. Then Equation (10) tells us that only those clusters with the smallest perimeter are important in averages over all cluster configurations. The smallest perimeter, for a cluster of $s = L^3$ sites on a simple cubic lattice, is obtained for configurations having no holes at all in their interior; their perimeter is $6L^2$ and their average radius is of the order of L. Thus R_s is proportional to $s^{1/3}$ for p close to unity, and not to $s^{\sigma v} = s^{0\cdot4}$ as for $p = p_c$. We have seen in Section 2.8 that one should expect the same type of asymptotic behaviour for all p above p_c (Equation (43b)). Thus we also expect this s law to be valid for the radius for all p larger than p_c: $1/\rho = 3$. In d dimensions we thus have

$$\rho = \frac{1}{d} \qquad (p > p_c) \tag{51}$$

which shows that ρ is not the same as at the threshold. Clusters above p_c are not fractals but 'normal' objects with $\rho = 1/d$.

Percolation theory supports equal rights for clusters above and below p_c; if above p_c they have the freedom to deviate from Equation (50), those below p_c have the same freedom. Again, one expects the same ρ for all p below p_c. In the limit $p \to 0$, all different perimeters in Equation (10) get the same weight, which means that the cluster radius is now the average radius of all animal configurations of the given size s. (All animals are equal, none of them are more equal than others.) Unfortunately, no general exact solution is known for animal radii, but in three dimensions we have $\rho = 1/2$ exactly (Parisi and Sourlas, as cited after Chapter 2). In two dimensions, ρ is about $0\cdot641$ from numerical estimates, and as mentioned above, $\rho = 1/4$ in the Bethe lattice. Thus the animals as well as the percolation clusters below p_c are again fractals, but with a fractal dimension $1/\rho$ smaller than that at the percolation threshold. Our table of exponents in the preceding chapter summarized the situation for $D = 1/\rho$.

Numerical studies of cluster radii have been made by exact evaluation of small clusters and by Monte Carlo simulations of larger ones. The results are in reasonable agreement with the above relations, as Figure 15 indicates.

No good numerical tests are known to me at present for the plausible scaling law

$$R_s = s^\rho h[(p - p_c)s^\sigma] \tag{52}$$

analogous to Equation (33). Just as the cluster numbers give different exponents ζ

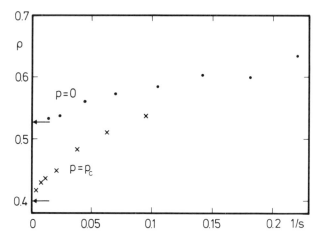

Figure 15. Monte Carlo data for the cluster radius exponent $1/D$ in the simple cubic lattice. We give the slopes of log-log plots of radius versus cluster mass s, as a function of $1/s$. The intercept of these graphs should give the asymptotic reciprocal fractal dimension D. These data are consistent with the theoretical expectations $1/D = 1/3, 0.4$ and $1/2$ above, at and below p_c for very large clusters and also show that for smaller clusters one has deviations.

and θ above, at and below p due to appropriate behaviour of the scaling function in Equation (33) (see end of Section 2.8), the cluster radii are also expected to have a scaling function h which behaves in such a way that it leads to different fractal dimensions above, at and below the threshold. Equations (33) and (52) are merely two examples for the general behaviour near critical points. Functions of two variables x and y (here $x = 1/s$ and $y = p - p_c$) can in most cases be approximated for small x and small y by functions of one variable z only, multiplied with a simple power of x; z equals y divided by some other simple power of x. In the case of Equations (33) and (52) we have $z = y/x^\sigma$; cluster numbers varied as $x^\tau f(z)$, and cluster radii presumably as $x^{-\rho}h(z)$. Indeed, if you write down any simple expression for a function of two variables x and y and if you avoid exponentials, logarithms or more complicated elements in this expression, then most of your attempts will lead to a result which for small x and small y has the simpler form

$$x^{-A}g(y/x^B)$$

of which Equations (33) and (52) are but two examples. You have to work quite hard to get some exceptions. Nature, apparently, was too lazy to work hard and instead followed these simple forms wherever it could in two and three dimensions for critical phenomena. (Should you dislike laziness you may also say that Nature has to obey $\xi = \infty$ right at the critical point. Therefore correct formulae at the critical point do not contain any large but finite length ξ.)

However, we have seen that Nature is not completely lazy; the radius exponent is not the same for all p (as in the Bethe lattice) but takes on three different values, valid above, at and below p_c. What does that mean for the structure of the clusters? If we define an average density as the ratio s/R_s^d of cluster mass to cluster volume, then above p_c with $R_s \propto s^\rho$ one has an average density independent of the cluster size for large clusters. This density equals the strength P of the infinite cluster since the interior of a very large cluster should not be different from that of the infinite network. Below and at p_c, the strength P of the infinite cluster is zero, and therefore the average density approaches zero if the cluster mass s goes to infinity. Franke has given the density profiles of large but finite clusters, that is the probability that a site at distance r from the cluster centre-of-mass belongs to that cluster. For p above p_c, the interior region of high densities near P is separated from the outside of the cluster (zero density profile) by a relatively narrow surface layer (Franke); below and at p_c this surface layer has spread over the whole interior of the cluster. Thus above p_c one has a rather narrow surface whereas below and at p_c the surface extends over the whole volume. Now it is no longer surprising that the exponent ζ for the asymptotic decay of cluster numbers (Equation (43)), has the surface value $(1 - 1/d)$ above p_c but is unity (surface proportional to volume) below p_c. Thus we see that the sites of finite clusters in percolation theory have some human traits. Above the threshold, the sites stay together like workers in a trade union and thus they achieve higher densities. Below the threshold they seem to prefer some distance from each other, scatter over a larger volume, the links between them can be broken more easily, just as with non-unionized people, and thus they achieve only low densities. In this sense you reach a percolation threshold if you enter a union.

3.3. The infinite cluster at the threshold

Is there or is there not an infinite cluster present at $p = p_c$? We know that there is one for p above the threshold, and there is none for p below the threshold. What does Nature do at the border line?

First we have to clarify what we mean by an infinite network, since real systems are always finite. We may call a cluster infinite if it connects the top line (top plane) with the bottom line (or plane). In this case, in a computer simulation of large lattices at $p = p_c$, a finite fraction, for example one-half, of all lattices have an infinite cluster in this sense, and the rest do not. Thus the answer to whether an infinite cluster is present is simply a 'perhaps'.

Instead we may look at the largest cluster in the finite system (without periodic boundary conditions). Of course, even for p far below p_c the system has a largest cluster. But only for p above p_c is the size of this largest cluster of the order of the system size; for p below p_c the size of the largest cluster increases only very weakly (logarithmically) with the system size. Now we ask: How does the size s of the largest cluster increase with L in a system with L^d sites? Below p_c it increases as $\log (L)$, above p_c as L^d; what happens at p_c? It would seem reasonable that the largest cluster will have a radius of the order of the system length L: $R_s \propto L$. Since at $p = p_c$ we have $R_s \propto s^\rho$, the condition for the largest size is simply $L \propto s^\rho$, very similar to Equations (48) and (49). Thus the infinite cluster at p_c (which sometimes is called the incipient infinite cluster) also is a fractal in our sense and has the same fractal dimension $1/\rho$ as large finite clusters at the threshold. Above p_c, the mass s of the infinite cluster increases as $L^{1/d}$ which means that it is no longer a fractal but $\rho = 1/d$ as for the finite clusters. Below p_c, the fractal dimension of the largest cluster is zero (corresponding to a very weak, that is logarithmic increase with L), in contrast to the fractal dimension of the finite clusters which are animal-like with $\rho = 1/2$ in three dimensions.

Figure 16 shows results in two dimensions for lattices containing up to 10^{10} sites. Apart from fluctuations we see a simple straight line in this plot of log s versus log L when L is not too small. The slope of this line is close to the theoretical value $1/\rho = 91/48$. Thus now we have a more quantitative answer to how large is 'infinite' for the incipient infinite cluster.

In this picture we took $p_c = 1/2$ as is known exactly for the triangular lattice. But even if p_c is not known exactly and if one takes a p_c slightly too high, one still observes for the largest cluster a mass s proportional to $L^{1/\rho}$, as long as L is much

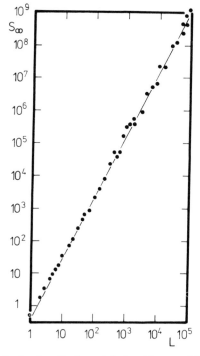

Figure 16. Monte Carlo data for the size of the largest cluster at the percolation threshold $p = p_c = 1/2$ of the triangular lattice, as function of the linear dimension L of the lattice. The slope of this log-log plot for large sizes gives the fractal dimension $D = 91/48$.

smaller than the correlation length ξ. (At the exact threshold, ξ is infinite, and this condition is always fulfilled). On the other hand, if L is much larger than ξ, the mass of the largest cluster will be PL^d, with P again being the strength of the infinite cluster, proportional to $(p - p_c)^\beta$. If L is of the order of ξ these two expressions PL^d and $L^{1/\rho}$ are also of the same order (assuming again that only one length ξ is dominating critical behaviour):

$$PL^d = const \ L^{1/\rho} \quad at \quad L = \xi \propto (p - p_c)^{1/\nu}$$

Thus

$$\beta - d\nu = -\nu/\rho$$

or

$$D = 1/\rho = d - \beta/\nu$$

Together with $\nu/\rho = 1/\sigma = \beta + \gamma$ and the other scaling laws we get

$$d\nu = \gamma + 2\beta = 2 - \alpha = \frac{(\tau - 1)}{\sigma} \tag{53}$$

Often this scaling law where the dimensionality d enters is called 'hyperscaling', and is also used for thermal phase transitions. It is obviously not correct for the Bethe lattice $(d = \infty)$ but seems good in two and three dimensions (also for thermal critical phenomena), which is what we are interested in.

Figure 2 has already shown us that the (incipient) infinite cluster has a very complicated structure. It does not seem to be a one-dimensional channel of occupied sites, but resembles more a network of roads, with many parking places and dead ends. Figure 17 gives a schematic picture along these lines, following Stanley and Coniglio (in the book on percolation structures and processes cited at the end of the Introduction). The one-dimensional channels or 'links' (roads) meet at many crossing points ('nodes') at distances of the order of ξ. Also, from these main roads many streets lead to dead ends ('dangling ends') where no through traffic can flow. Finally, on the main roads one sometimes finds large parking lots ('blobs') where many more or less parallel roads can carry traffic. Links, nodes and blobs together form the backbone of the infinite network, whereas the dead ends only increase the total mass of the infinite cluster without helping it to span the whole system.

More quantitatively, we may imagine an electric voltage to be applied to the top and the bottom line (plane) of our system, as also indicated in Figure 17. Occupied sites are conductors, empty sites cannot carry any electric current. Then the dead ends carry no electric current whereas the backbone does. Most of the mass sits in dead ends. Within the backbone the dangerous or 'red' parts are those which cannot be removed without bringing the total conductivity to zero. If one of the other ('blue') backbone sites is removed the current still can flow

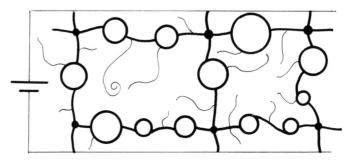

Figure 17. Schematic picture for the links (one-dimensional chains), nodes (crossing points of the links) and blobs (dense regions with more than one connection between two points; shown as circular here) of the infinite cluster slightly above the threshold. The distance between the nodes as well as the blob diameter are assumed to be of the order of the correlation length. The thin lines are the dead ends; for clarity only very few of them are shown. Most of the material is in the dead ends, the rest is called the backbone. Most of the backbone mass belongs to blobs.

elsewhere. The blobs therefore mainly consist of blue sites, not of red ones. Numerical studies were made for example by Pike and Stanley (1981) on the average number of different types of network parts. A simple power law $L^{1/\nu}$ was proven by Coniglio for the number of red bonds in a section of size L^d of the incipient cluster. These details are beyond the scope of this book and the author's competence.

A different attempt to approximate the incipient infinite cluster at $p = p_c$ is the use of deterministic fractals to describe the random percolation clusters. We learned above that fractals have a mass increasing with the D-th power of their radius. Perhaps one can construct other objects, which are completely regular, can be studied exactly (not by computers), and have the same fractal dimension as the incipient infinite cluster ($1/\rho = 91/48$ in two dimensions). One of these fractals is shown in Figure 18 and is called the Sierpinski carpet. It is constructed as follows. Let us start with an occupied square: its mass s is 1, and so is its length L. Now the occupied square is replaced by 3×3 squares, of which the centre square is empty, the neighbouring eight squares are occupied, as shown in Figure 18. The mass now is 8, not counting the empty squares, and $L = 3$. This process is repeated again and again, with occupied squares replaced by 8 occupied and one empty centre square, and empty squares replaced by 9 empty squares. Figure 18 shows the next step with $s = 64$ occupied squares and a length $L = 9$. At each iteration L increases by a factor 3 and s increases by a factor 8. After n iterations we thus have

$$L = 3^n \qquad s = 8^n$$

thus

$$s = L^D$$

where

$$D = \frac{\log 8}{\log 3}$$

n=0

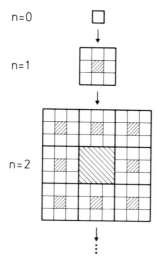

n=1

n=2

Figure 18. Initial stages of the build-up of the Sierpinski carpet, a simplified fractal as an approximation of the incipient infinite cluster. Empty squares are shadowed. At each step of the iteration, the linear dimension L is enlarged by a factor 3, and the mass by a factor 8, since each occupied square is replaced by a 3×3 array of 9 squares of which the centre square is empty.

This fractal dimension $D = 1 \cdot 893$ is nearly exactly equal to the $1/\rho = 1 \cdot 896$ for the incipient infinite cluster of percolation in two dimensions. Other properties can also be studied and sometimes good agreement with percolation properties can be found. However, it seems difficult to produce simple objects which show the typical percolation behaviour as a function of a parameter p; that is one exponent for all p above p_c, another for all p below p_c, and a third one (the above fractal dimension) only at $p = p_c$. We refer to Mandelbrot (1982 and 1984) or Aharony *et al.* (1984) for more details.

Further reading

Aharony, A., Gefen, Y. and Kantor, Y., *J. Statist. Phys.*, **36**, 795, (1984) and other papers at this fractal conference.

Coniglio, A., *Phys. Rev. Letters*, **46**, 250, (1981).

Franke, H., *Z. Physik B*, **40**, 61, (1981); **45**, 247, (1982), *Phys. Rev. B*, **25**, 2040, (1982).

Franke, H. and Kertész, J., *Phys. Letters*, **95**, 52, (1983).

Kapitulnik, A., Aharony, A., Deutscher G. and Stauffer, D., *J. Phys. A*, **16**, L 269, (1984).

Mandelbrot, B.B., *The Fractal Geometry of Nature* (San Francisco: Freeman, 1982) and *J. Statist. Phys.*, **34**, 895, (1984).

Pike, R.B. and Stanley, H.E., *J. Phys. A*, **14**, L 169, (1981).

Zimm, B.H. and Stockmayer, W.H., *J. Chem. Phys.*, **17**, 1301, (1949).

CHAPTER 4

percolative renormalization group

In 1971, K.G. Wilson published the first renormalization group treatment of critical phenomena and was honoured a decade later by the Nobel prize for physics (though in the cumulative author index of the journal at that time the articles were forgotten). It is an attempt to justify the scaling assumptions made earlier, and to calculate the critical exponents entering these scaling assumptions. Historically, it was first applied to thermal phase transitions and only afterwards to percolation; also initially it dealt with fluctuations in Fourier space (as function of wave vector) and only later moved to real space (where everything depends on distances). Ignoring this history, we will concentrate on real space renormalization of percolation (sometimes also called position space renormalization). This seems the simplest way to introduce renormalization ideas into percolation theory (Reynolds *et al.*, 1980) since the method then becomes more or less equivalent to finite-size scaling. Thus we will first explain finite-size scaling and then go to renormalization techniques. For thermal critical phenomena, real space renormalization is usually more complicated, though some examples exist (Binder (1981) for static, Jan *et al.* (1983) for dynamic properties) which are also closely related to finite-size scaling.

4.1. Finite-size scaling

How do the various quantities of interest behave near the percolation threshold in a large but finite lattice? Let us take, as an example, the normalized size of the largest cluster, that is the probability P that an arbitrarily selected site belongs to the largest cluster of the system.

For an infinitely large system, this largest cluster is infinitely large above the percolation threshold and thus, if its size is divided by the system size, the ratio P is finite' $P(p > p_c) > 0$. For concentrations p below p_c, even the largest cluster is relatively small and the ratio of its size to the lattice size goes to zero if the lattice size goes to infinity, $P(p < p_c) = 0$. For p slightly above p_c we learned in Equation

(29) that $P \propto (p - p_c)^\beta$. How are these results changed in a finite system when P depends not only on the concentration p but also on the linear dimension L of the lattice, for example in a $L \times L \times L$ simple cubic lattice.

We have already learned after Equation (52) how scaling theory deals with functions like $P(p - p_c, L)$ depending on two variables. In the asymptotic regime, that means large lattices ($1/L$ near zero) and close to the threshold ($p - p_c$ close to zero), we expect P to follow

$$P = L^{-A} F[(p - p_c)L^B]$$

from the mathematical analogy with Equation (52) or Equation (33). Here F is a suitable scaling function and A and B are suitable critical exponents. Since for a finite system the largest cluster always has a finite size, which moreover does not change drastically near the threshold, the scaling function $F(z)$ will always be a positive function of its argument $z = (p - p_c)L^B$ and is analytic everywhere, even at the percolation threshold where $z = 0$. For $L \to \infty$ at fixed p above p_c we must recover $P \propto (p - p_c)^\beta$ independent of L. Thus for very large positive z, the scaling function $F(z)$ must vary as $z^{A/B}$ in order that L cancels out. Then we get $P \propto (p - p_c)^{A/B}$ and thus $A/B = \beta$. On the other hand, for $z = 0$, that is at the percolation threshold, the largest cluster will have a radius R_s of the order of the system length L. From Equation (49) we therefore find the largest cluster to contain about $L^{1/\rho}$ sites at $p = p_c$. The probability P (size divided by lattice size) thus varies as $L^{-d + 1/\rho}$ in d dimensions. The above assumption, on the other hand, gives $P = L^{-A} f(0)$ at the threshold. Thus $A = d - 1/\rho = \beta/\nu$, using the relation mentioned before Equation (53). Our above relation $A/B = 1/\nu$ now gives us $B = A/\beta = 1/\nu$ and

$$P = L^{-\beta/\nu} F[(p - p_c)L^{1/\nu}] \tag{54}$$

Similarly, for any other quantity varying as $|p - p_c|^x$, we can also apply Equation (54), with β replaced by the appropriate exponent x, and with a different shape of the scaling function $F(z)$, but with the same correlation length exponent ν of Equation (47). At $p = p_c$, this quantity then varies as $L^{-x/\nu}$. This result gives an accurate way to determine critical exponents. As already mentioned, Figure 16 determines the fractal dimension $1/\rho = d - \beta/\nu$ and thus β/ν. If the mean cluster size or second moment of the cluster size distribution at $p = p_c$ is plotted versus system size, it varies as $L^{\gamma/\nu}$, etc. If the exponent ν is known from other methods (see below), we therefore can determine β and γ.

What does Equation (54) really mean? Its simplicity requires that there be only one correlation length $\varepsilon \propto |p - p_c|^{-\nu}$ in our system; otherwise two different exponents ν could appear in Equation (54). (That may be the case for $d > 6$.) If this correlation length ξ is much smaller than the system length L, that is if $p - p_c \gg L^{-1/\nu}$, we do not see the boundaries of the system in our simulations and quantities averaged over the whole volume behave as in an infinitely large system.

This is the case if the argument $|z|$ is much larger than unity. If we get closer to the threshold at a fixed finite system size L, the correlation length ξ increases and the argument z decreases. When ξ has reached a value of the order of L, $|z|$ will be of the order of unity, and then appreciable deviations from the behaviour of infinite systems are expected. If we get even closer to p_c, z will be nearly zero, and the behaviour is nearly that observed right at $p = p_c$. This general behaviour is depicted schematically in Figure 19 for the strength P of the 'infinite' network, as described mathematically by Equation (54). Of course, all these scaling relations are only valid for rather large lattices and close to the percolation threshold; one cannot learn much about percolation theory by looking at a single occupied site.

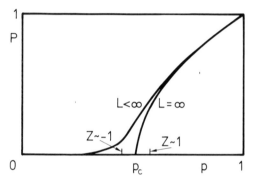

Figure 19. Schematic behaviour of the percolation probability P according to finite-size scaling theory, Equation (54). For $|z| \gg 1$ one has bulk behaviour even in a system of finite length L whereas for $|z| \ll 1$ one has $P \propto L^{-\beta/\nu}$. The border $|z| = 1$ corresponds to correlations lengths ξ of the order of L.

Let us now apply these ideas to the probability $R = R(p, L)$ that a lattice of linear dimension L percolates at concentration p. (We define a lattice as percolating if at least one cluster connects the top line or plane with the bottom line or plane.) In an infinite system, we have $R = 1$ above and $R = 0$ below p_c. The quantity dR/dp gives the probability (divided by the small interval dp), that the lattice starts to percolate if the concentration is increased from p to $p + dp$. Since in infinite systems $R = 1$ for all p above p_c, the critical exponent of R is zero, and the analogue of Equation (54) in a finite system is

$$R = \Phi\left[(p - p_c)L^{1/\nu}\right] \tag{55a}$$

for large L close to p_c. The scaling function Φ increases from 0 to 1 if its argument increases from $-\infty$ (far below threshold) to $+\infty$ (far above threshold). The derivative gives

$$\frac{dR}{dp} = L^{1/\nu}\Phi'\left[(p - p_c)L^{1/\nu}\right] \tag{55b}$$

(For $L \to \infty$, this derivative approaches a delta function, as physics students call it, or a delta distribution as mathematicians insist on calling it. Ordinary people simply call it a very narrow and high peak.) Figure 20 shows schematically how R and dR/dp behave for medium and large lattices, in agreement with numerical studies (see, for example, the review of Clerc $et\ al.$ (1983) mentioned after the Introduction).

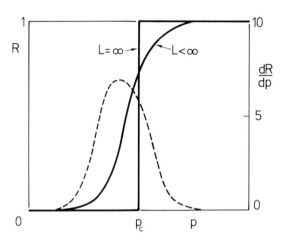

Figure 20. Variation of the probability R (solid lines) that a cluster is spanning the whole system, for medium and large system sizes. The dashed lines give dR/dp, proportional to the probability that at concentration p a spanning cluster starts to appear. The width of the transition region or peak varies as $L^{1/\nu}$. (Schematic.)

The average concentration p_{av} at which, for the first time, a percolating cluster connects top and bottom of the cluster is defined as

$$p_{av} = \int p \left(\frac{dR}{dp} \right) dp \qquad (56)$$

the integral here and later runs from $p=0$ to $p=1$. (Note $\int (dR/dp)\,dp = R(1) - R(0) = 1$.) Since dR/dp is basically the probability that at concentration p such a spanning cluster appears for the first time, we can determine p_{av} by making numerous Monte Carlo experiments for the same L and check when the system percolates for the first time if we slowly fill up the sites. For large lattices it is practical to achieve this by the following approximation. First take $p=1/2$ and check if the lattice percolates. If it does, decrease p by $1/4$, otherwise increase p by $1/4$. Then check again if the system percolates; decrease p by $1/8$ if it does and decrease p by $1/8$ if it does not. Repeat this division until one has determined with sufficient accuracy the concentration at which the first spanning cluster appears. After every change of p the random number generator has to be reset at its original value in order that in the following simulation most of the sites previously occupied (empty) will be occupied (empty) again. After about ten such iterations the onset of percolation is known with an accuracy sufficient for many purposes, but this value is only the value for this particular sequence of random numbers. Now we have to repeat the Monte Carlo simulation again and again and get, in general, each time a different concentration at which for the first time a cluster connects top with bottom. Averaging the observed onset concentration over all these different sequences of simulations we get an estimate for p_{av}. (Instead of checking for a spanning cluster one may also determine a different percolation threshold as the position of the maximum in the second moment of the cluster size distribution (Jan *et al.*, 1984). For large systems these thresholds converge to the same limit p_c as those described here.)

How does this effective percolation threshold p_{av} for one system size L approach the asymptotic value p_c for infinite systems? From Equations (55) and (56) we find

$$p_{av} - p_c \propto L^{-1/\nu} \tag{57}$$

with the proportionality constant being $\int z\Phi'(z)\,dz$. (In special cases like the triangular lattice the proportionality constant may be zero since then dR/dp is completely symmetric about $z = 0$.) This variation of p_{av} with system size L is one way to determine the critical exponent ν: one plots p_{av} versus $L^{-1/\nu}$ for various trial values of ν and selects that value for ν which gives the best straight line for large L. (If one has lots of data and a suitable fitting program one can also determine ν, β and p_c as those parameters which best fulfil the finite-size scaling assumption of Equation (54).)

Not only p_{av} approaches p_c as $L^{-1/\nu}$. If we also define an effective percolation threshold for finite lattices as that point where the curve $P(p)$ has an inflexion point (maximum of dP/dp), this L-dependent threshold approaches the true p_c as $L^{-1/\nu}$, as Equation (54) tells us immediately. By analogy, the maximum of the mean cluster size $S(p)$ gives us an effective threshold value which approaches the true threshold as $L^{-1/\nu}$. In other words, Equation (57) is valid for every reasonable definition of a percolation threshold for finite large systems, not just for p_{av}. Only the proportionality constant is different for different definitions of the onset of percolation.

We mentioned above that the exponent β/ν can be determined from $P \propto L^{-\beta/\nu}$ by simulating a system exactly at $p = p_c$ for different L. We now see that we may also simulate it at some suitably defined size-dependent threshold, like p_{av}. For then the argument $z = (p - p_c)L^{1/\nu}$ in Equation (54) is a constant, instead of being zero, and the proportionality for $P(L)$ is not affected.

All these remarks are also valid for thermal critical phenomena in two and three dimensions (Fisher, 1971) and were used there successfully before they were applied to percolation. One has only to replace P by the magnetization, and $p - p_c$ by $T_c - T$, if one deals with the ferromagnetic Curie point.

The width Δ of the transition region between small and large probabilities R of a spanning cluster can be defined very simply as the difference between the concentration where R is 0·1 and the concentration where R is 0·9. One may also take 0·2 and 0·8, or 0·1 and 0·8, as suitable numbers to be used in the definition of this width. Equation (55a) then tells us that this width Δ, independent of the details of its definition, varies for large L as $L^{-1/\nu}$. A less arbitrary way to define the width is

$$\Delta^2 = \int (p - p_{av})^2 \left(\frac{dR}{dp}\right) dp \qquad (58a)$$

Thus Δ is the root mean square deviation of the thresholds actually observed, from their average value. For averages $\langle x \rangle$ of fluctuating quantities x, one can easily show quite generally that the average of $(x - \langle x \rangle)^2$ equals $\langle x^2 \rangle - \langle x \rangle^2$. Thus one can easily determine Δ from a series of Monte Carlo evaluations of the actual onset p of percolation by summing the values of p as well as p^2 for each sequence of random numbers. At the end one can then calculate

$$\Delta^2 = \langle p^2 \rangle - \langle p \rangle^2$$

and determine the correlation length exponent ν from

$$\Delta \propto L^{-1/\nu} \qquad (58b)$$

This method works particularly well for percolation where it was introduced by Levinshtein *et al.* (1975); for thermal phase transitions one does not in general have as simple a definition for an effective critical point as the onset of a spanning cluster for percolation, and thus Equation (58b) has seldom been applied there. We will mention numerical results in the next section, where we will show the close connection with renormalization group techniques. Here we simply warn the reader that one needs at least hundreds of simulations to get the exponent ν accurately from Equation (58), whereas far fewer are needed to estimate p_c reliably from Equation (57). Thus determinations of the threshold should be done with relatively few simulations of large lattices whereas one needs many more simulations for the determination of the correlation length exponent. Therefore because of demands on computer time these have to be done for smaller systems.

4.2. Small cell renormalization

The basic idea of renormalization is self-similarity at the critical point. What does that mean? We saw in Equation (54) that the crucial question of finite-size scaling is the question whether the system length L is larger or smaller than the correlation length $\xi \propto |p - p_c|^{-\nu}$. We saw in Equations (33) and (52) that clusters can be separated into two main groups: those with mass s larger than $|p - p_c|^{-1/\sigma}$, and those with smaller s. For small clusters or small systems, one kind of power law is valid, for example $P \propto L^{-\beta/\nu}$, and for large clusters or systems another power law holds, for example $P \propto (p - p_c)^\beta$. In other words, all clusters or systems smaller (in linear dimension) than the correlation length ξ are similar to each other in an averaged sense, as long as they contain many sites. This similarity breaks down for large sizes of the order of ξ as well as for small sizes of the order of the distance a between nearest neighbours on the lattice. Right at the percolation threshold the correlation length is infinite, thus all large clusters or lattices are similar to each other. If we take out a medium size part of a bigger lattice, then both this part and the bigger lattice are still much smaller than ξ at $p = p_c$, and thus similar to each other in an average sense.

I refrain from showing you a picture purporting to prove to you that similarity holds for distances between a and ξ, for a critical reader would not really find such a similarity. Only if one averages over many pictures can a computer find such similarity. I regard Figure 16 as a much better 'proof'. If different lattice sizes were not similar to each other at the threshold one should not be able to observe a simple power law as seen by the straight line in Figure 16. For a completely regular fractal like the Sierpinski carpet of Figure 17, this self-similarity holds already without any averages. The reader will find many more examples of geometrical similarity in Mandelbrot's book (cited in the preceding chapter). The similarity idea as a foundation of thermal critical phenomena and scaling goes back to the 1960s (see the review of Kadanoff *et al.* 1967) and lead to Wilson's first renormalization theory.

In real space renormalization, we replace a cell of sites by a single super-site, provided that the linear dimension b of the cell is much smaller than ξ. Of course we lose information if, say, 16 sites of a 4×4 cell in the square lattice are replaced by a single super-site. But if scaling relies on the fact that all cells of size b are similar to each other, then perhaps we should get a good critical exponent out from this approximation where we renormalize a whole cell of, say, b^d sites into a single super-site. Quantitatively, such a renormalization of cells to sites requires a certain rule governing how this is to be done; moreover the concentration p' of occupied super-sites will in general be different from that of the original sites. Only right at the critical point, where self-similarity is valid, do we have $p' = p = p_c$. In general we know that the correlation length ξ limits the validity of similarity; thus this limit ξ is the same in both the original lattice and the renormalized lattice of super-sites, $\xi = \xi'$. If in the original lattice we have $\xi = const\, |p - p_c|^{-\nu}$ then in the renormalized lattice, with lattice constant b, we have $\xi' = const\, b|p' - p_c|^{-\nu}$, with the same proportionality constant and the same critical exponent ν. Thus

$$b|p' - p_c|^{-\nu} = |p - p_c|^{-\nu} \tag{59a}$$

is the basic equation of real space renormalization. Taking the logarithm of both sides we arrive at

$$\frac{1}{v} = \log \left[\frac{(p'-p_c)}{(p-p_c)} \right] \Big/ \log (b)$$

(59b)

for the exponent of the correlation length; often $1/v$ is abbreviated as y or y_T or y_p in renormalization publications. In summary, we renormalize a cell of size b into a single super-site; to keep the real quantity ξ unchanged in this renormalization, we also have to renormalize p into p'.

The above ideas also apply to thermal critical phenomena as well as to percolation (with changed notation, of course; p then is replaced by the ratio of interaction energy to thermal energy). The basic difference lies in the rule governing how to renormalize a given cell provided that we know the status of every site within the cell. For thermal critical phenomena, like the renormalization of magnetic spins, it seems sensible to apply a majority rule. If in a 5×5 cell on the square lattice at least 13 of the 25 spins point upwards, then the super-spin representing this cell points upwards; otherwise it is defined as pointing downwards. In percolation, however, we are less interested in the total number of occupied sites than in the way they are connected (Reynolds *et al.*, 1980). Therefore we define the super-site representing the cell of length b as occupied if the cell contains a spinning cluster connecting opposite lines or planes, otherwise the super-site is defined as empty. In other words, thermal critical phenomena correspond to a democratic society where you need majority support whereas percolation is similar to a corrupt society where it counts whom you know closely and how well you are connected. Don't blame me for it; I did not invent renormalization nor was I the first to apply it to percolation (Harris *et al.*).

Whether you like it or not, corruption works. Let us take as the simplest example the triangular lattice. Each triangle contains three sites at its corners, and we place the super-site into the centre of the triangle. Figure 21 shows that we only renormalize suitable triangles in such a way that each original site belongs to exactly one triangle. We now ask for the probability p' of such a super-site belonging to a renormalized triangle, being occupied if every original site is occupied with probability p. The super-site is occupied if a spanning cluster exists. In our triangle this is the case if either all three sites are occupied (probability p^3), or if two neighbouring sites are occupied and thus connect two opposite ends of the triangle. The latter case can be realized in three ways (depending on which of the three possible pairs is occupied), each of which has the probability $p^2(1-p)$, see Figure 21. Thus our renormalized probability is

$$p' = p^3 + 3p^2(1-p)$$

(60)

A plot of this function $p'(p)$ has some similarity with the function $R(p)$ depicted in Figure 20. (To support more ethics in government I selected an example where

Figure 21. Real space renormalization of a triangular lattice. The circles denote the super-sites, each representing three different original sites. The super-sites again form a triangular lattice.

majority rule and connectivity rule give identical results. So at least I pretend not to be corrupt.)

Right at the critical point we should have complete similarity: $p' = p$. We call the concentration p^* where this condition is valid the 'fixed point', also a crucial concept for more complicated renormalization problems. The renormalization leaves fixed points fixed. In our case, the equation $p' = p$, with p' from Equation (60), has three solutions:

$$p^* = 0, \ 1/2 \ \text{and} \ 1$$

The first (zero) and the last (unity) solution are quite trivial and exist also for different lattices and dimensionalities; we are interested only in the non-trivial fixed point $p^* = 1/2$. This fixed point agrees exactly with the known critical point p_c of the triangular lattice, a first indication that the renormalization idea might be correct, after all. Now we expand Equation (60) about this fixed point:

$$p' = p^* + \lambda(p - p^*) + 0(p - p^*)^2$$

with

$$\lambda = dp'/dp = 6p - 6p^2 = 3/2 \qquad \text{at} \ p = p^* = 1/2$$

Equation (59b) now has the form

$$\frac{1}{v} = \frac{\log \Delta}{\log b} \tag{61}$$

In our particular lattice we have $b^2 = 3$ since in the plane three old sites form one super-site (see Figure 20). Thus

$$v = \frac{\log (3^{1/2})}{\log (3/2)} = 1 \cdot 355$$

This result is an excellent approximation to the presumably exact $v = 4/3$ in two dimensions, confirming that renormalization group works.

Unfortunately, this excellent agreement for v and p_c is rather exceptional. For other lattices or in other dimensions there are usually stronger deviations. Only if we let b go to infinity can be expect that the renormalization result will approach the exact value. In a square lattice with $b = 5$, renormalization of the 25 sites in a 5×5 cell has to deal with 2^{25} different configurations, not an easy task (Reynolds *et al.*, 1980). To go to larger systems we use a Monte Carlo simulation to deal with a representative sample of all possible configurations. This method will be described in the following Section 4.3.

Another way to look at large cells is to evaluate the behaviour of two-dimensional strips of width b and infinite length. When b is of the order of ten one can still do that exactly; for larger b one again needs Monte Carlo simulation. The main disadvantage is that this method is mainly restricted to two dimensions if one wants to avoid Monte Carlo simulation. For in three dimensions, an infinite bar of cross-section $b \times b$ already contains b^2 sites in each plane, which is easily more than the ten to twenty sites which can still be handled exactly. On the other hand, for two dimensions it gives very accurate estimates. It was crucial to have these estimates in order to believe the supposedly exact critical exponent of two-dimensional percolation in Table 2, Section 2.7. A review of this method was given by Vannimenus and Nadal (1984). Since much of this work on strips was done in Paris, the method can be called striptease, though names like 'phenomenological renormalization', 'Nightingale renormalization', 'transfer matrix approach', or just finite-size scaling, look more scientific.

An advantage of these renormalization ideas is the qualitative information they give in addition to the position of the fixed point and the critical exponents. If in Equation (60) we take a concentration $p = p_c - \varepsilon$ slightly below the fixed point $p^* = p_c$, the renormalized concentration $p' = p_c - 3\varepsilon/2$ is even further away. We may apply the renormalization rule (60) to this new concentration p' and get a third concentration p'' lower than the previous ones. Each of these transformations corresponds to the replacement of three sites (which may be super-super-sites etc.) by one new site. When we have applied this renormalization very often, the concentrations will be close to zero, and then in our case each new concentration is roughly three times the square of the old one and thus even smaller than the old one. The renormalization equation (60) therefore flows into the trivial fixed point $p^* = 0$, if we start with a p slightly below 1/2. Starting with p slightly above 1/2, we flow into the other trivial fixed point $p^* = 1$ if we apply the renormalization of Equation (60) again and again. These flows to zero and unity suggest that there is no other critical point between the usual percolation threshold p_c and the 'animal' fixed point $p = 0$. Therefore all clusters have the asymptotic critical exponents $\zeta = 1$ and θ of lattice animals, if $p < p_c$, Section 2.8. (Asymptotically large cluster sizes may be identified with very many iterations of Equation (60).) On the other hand, all concentrations above p_c flow to the 'compact' fixed point $p^* = 1$, which means that their asymptotic exponents like $\zeta = 1 - 1/d$ are the same for all p above p_c.

To get other critical exponents besides v and to deal with more complicated problems, one introduces additional variables besides p and renormalizes them. Linearization of the transformation formulae about the fixed point then gives a matrix with eigenvalues λ, related to critical exponents via equations similar to Equation (59b). Flow diagrams show us to which fixed points we move, and which systems should have the same critical exponents. The principles are, however, the same as those explained above.

4.3. Monte Carlo renormalization

We have mentioned already in the preceding section on small cell renormalization that small cells are not good enough. Similarity and power laws only hold asymptotically, that is for lengths much larger than the lattice constant. Thus good cells for renormalization must contain many sites, not just a few. In general, square or cubic cells with numerous sites cannot be renormalized by exactly known formulae. Thus we simulate randomly occupied cells on the computer by our well known Monte Carlo method, and then renormalize them.

How do we make the renormalization? We check whether the now rather large cell percolates, that is whether it is spanned by a cluster connecting top and bottom. This, however, was just what we described in the finite-size scaling Section 4.1 to find p_{av} and the width Δ. Thus the computer simulation is the same as in ordinary Monte Carlo work, only the analysis may differ. Similarly, for thermal critical phenomena the Monte Carlo renormalization technique first requires an ordinary simulation of the system; only the analysis of the resulting configurations may be different from finite-size scaling.

For percolation we thus see that the renormalized cell occupation probability p' is nothing but the spanning probability R. Its derivative dp'/dp shows a peak near the average percolation threshold p_{av}, which in turn approaches the true percolation threshold p_c if the cell size goes to infinity, Equation (57). The width Δ of the peak in $dp'/dp = dR/dp$ vanishes as $b^{-1/v}$, where now the cell size b of renormalization theory replaces the lattice size L of finite-size scaling (a mere change in notation; we still simulate one $b \times b$ square lattice at a time, for example). For every finite cell, the function $p' = R(p)$ is a polynomial in p, as we saw in the above example of the triangle, Equation (60). For $b \to \infty$, the peak is very narrow, and as for many other peaks in nature it approaches a

Gaussian. [A Gaussian $f(z)$ is a function proportional to exp $(-z^2/2 \, \Delta^2)$, where Δ is the width and obeys

$$\Delta^2 = \frac{\int z^2 f(z) \, dz}{\int f(z) \, dz}$$

The integral over this exponential function gives $(2\pi)^{1/2} \, \Delta$, with our integrals running from $-\infty$ to $+\infty$.] Similarly, the probability that M out of N different sites are occupied, if each site is occupied with probability p, is a polynomial in p involving the binomial coefficients; but if N goes to infinity it approaches a Gaussian, as a function of M, with a width proportional to $N^{1/2}$. In our case, what else could dR/dp do but be Gaussian for large cells? Thus we have

$$\frac{dR}{dp} = (2\pi)^{-1/2} \, \Delta^{-1} \exp\left[\frac{-(p - p_{av})^2}{2 \, \Delta^2} \right] \tag{62}$$

where we determined the prefactor from the condition that the integral over the probability dR/dp must be unity. The width Δ as well as $p_{av} - p_c$ varies as $b^{-1/\nu}$, as explained in Section 4.1.

Now let us apply our renormalization ideas. If p_c equals $1/2$, then p_{av} may also equal $1/2$ even for small systems. This is indeed the case for suitable triangular site or square bond percolation cells. If p_c differs from $1/2$, then p_{av} for finite cells differs from the p_c of infinite systems by an amount varying with cell size b as $b^{-1/\nu}$, see Equation (57). A similar relation holds for the fixed point p^* defined through $R(p^*) = p^*$ or

$$p^* = \int \left(\frac{dR}{dp} \right) dp$$

where the integral runs from $p = 0$ to $p = p^*$. Our Gaussian approximation gives

$$p^* = (2\pi)^{-1/2} \int \exp \left(-z^2/2 \right) dz$$

with $z = (p - p_{av})/\Delta$ and with this integral running from $z = -\infty$ to $z = (p^* - p_{av})/\Delta$. This second integral can only give a constant value as required, if its upper cutoff is constant, that is if $p^* - p_{av}$ varies as Δ, that is as $L^{-1/\nu}$. Thus we have shown, to no surprise, that all differences $p_{av} - p_c$, $p^* - p_c$, $p^* - p_{av}$ vary as the width Δ of the distribution. What else should they do if they obey finite-size scaling? We assumed that there is only one width, and got out what we assumed. (We did not even use fully the fact that the probability distribution dR/dp is Gaussian.)

However, in Section 4.2 we have already mentioned that shifts of effective critical points, whether p^* or p_{av}, are not the best way to determine the correlation length exponent v. What instead do Equations (59b) and (61) tell us now? We again expand linearly about the fixed point: $p' - p^* = \lambda(p - p_c)$ with $\lambda = dR/dp$ at p^*; Equation (62) tells us

$$\lambda = (2\pi)^{-1/2} \, \Delta^{-1} \exp(-const)$$

with $const = (p^* - p_{av})^2/2 \, \Delta^2$ which is independent of cell size according to what we just found. Thus from Equation (61) we get

$$y(b) = \frac{\ln(1/\Delta)}{\ln(b)} - \frac{const'}{\ln(b)}$$

with ln in the natural logarithm and $const' = const + \ln(2\pi)/2$. This result gives us, for every cell size b, an exponent $y = y(b)$. For $b \to \infty$ this effective exponent $y(b)$ should approach the true exponent $1/v$, since renormalization works exactly only for very large cells. Indeed, since Δ varies as $b^{1/v}$ asymptotically, the above expression is compatible with this requirement; it shows us moreover that the effective $y(b)$ for finite cells differs from the asymptotic exponent $1/v = y(\infty)$ by, among other terms, a term proportional to $1/\log(b)$.

Thus how can we in practice determine the asymptotic exponent $1/v$ from our widths $\Delta(b)$ for finite cell sizes b? We plot the ratio $\log(1/\Delta)/\log(b)$ versus $1/\log(b)$ and look for a smooth curve fitting these data and having a finite slope for $1/\log(b)$ going to zero. Instead of Δ one may also use $(2\pi)^{1/2} \, \Delta$ as argument of the logarithm to follow more closely the above derivation. In some examples the ratios $\log(1/\Delta)/\log(b)$ are larger than the extrapolated value $1/v$, in some cases they are smaller, but usually they can be fitted reasonably well on smooth curves with a finite slope at the intercept. Figure 22 gives an example for the triangular lattice.

What does that fitting procedure lead us to? If we look only at large b where the data are fitted reasonably well by the tangent on the curve with slope C and intercept $1/v$, we simply have

$$\frac{\ln(1/\Delta)}{\ln(b)} = \frac{1}{v} + \frac{C}{\ln(b)}$$

or

$$\Delta = \exp(-C)b^{-1/v}$$

This result, on the other hand, is nothing but our finite-size scaling result of Equation (58b) for the widths of the threshold distributions. Thus instead of going through the above analysis with $\log(1/\Delta)/\log(b)$ one may also simply take a log-log plot of widths Δ versus cell size b or L and determine $y = 1/v$ from the asymptotic slope of that plot. (For fitting techniques see the Appendix.) In this

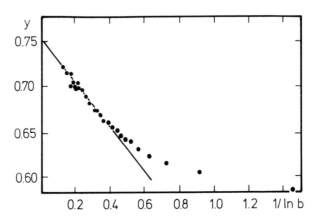

Figure 22. Results of large cell renormalization for triangular lattice, using b up to 10000 (Eschbach *et al.* 1981). The b-dependent effective exponents y, determined from the width of the distribution function for the threshold, are plotted versus $1/\ln(b)$ (full dots). A tangent on the values for large b has the 'true' $y = 1/v$ for infinite systems as intercept. These data are compatible with the intercept being 0.75, corresponding to the supposedly exact $v = 4/3$.

sense, Monte Carlo renormalization of large cells is equivalent to finite size scaling, if we look at these percolation problems. We mentioned already that for thermal critical phenomena such simple relations do not hold in general. Thus again percolation works also as a particularly simple way to enter into modern phase transition research.

Further reading

Binder, K., *Z. Physik B*, **43**, 119, (1981).
See also Stauffer, D., *J. Phys. A*, **17**, L 925 (1984).
Eschbach, P.D., Stauffer, D. and Herrmann, H.J., *Phys. Rev. B*, **23**, 422, (1981).
Fisher, M.E., in: *Proceedings of the Intern. Summer School Enrico Fermi, Course 51 Critical Phenomena*, Varenna, Italy, (New York: Academic Press, 1971).
Harris, A.B., Lubensky, T.C., Holcomb, W.K. and Dasgupta, C., *Phys. Rev. Letters*, **35**, 327, (1975).
Jan, N., Moseley, L.L. and Stauffer, D., *J. Statist. Phys.*, **33**, 1, (1983). See also Kalle, C., *J. Phys. A*, **17**, L 801, (1984).
Jan, N., Hunter, D.L. and Lookman, T., *Phys. Rev. B*, **29**, 6356, (1984).
Kadanoff, L.P. *et al.*, *Rev. Mod. Phys.*, **39**, 395, (1967).
Levinshtein, M.E., Shklovskii, B.I., Sur, M.S. and Efros, A.L., *Zh. Eksp. Teor. Fiz.*, **69**, 386, (1975); English translation: *Soviet Phys. JETP*, **42**, 197, (1976).
Reynolds, P.J., Stanley, H.E. and Klein, W., *Phys. Rev. B*, **21**, 1223, (1980).
Vannimenus, J. and Nadal, J.P., *Phys. Reports*, **103**, 47, (1984).
Wilson, K.G., *Phys. Rev. B*, **4**, 3174 and 3184, (1971).

CHAPTER 5

conductivity and other kinetic aspects

If you have studied equilibrium thermodynamics and/or statistical physics, and had some time left, you might have encountered the problem of non-equilibrium effects. For example, in equilibrium thermodynamics the temperatures of two objects in thermal contact, like Scotch and ice cubes, are the same. In non-equilibrium thermodynamics, we learn how quickly heat can diffuse from the warmer to the colder object, that is how long it takes to establish equilibrium. Similarly, in the scaling theory of phase transitions, shortly after static scaling laws were invented during the sixties, they were generalized to cover time-dependent or non-equilibrium effects like the thermal conductivity near the superfluid transition or the spin wave spectrum near the ferromagnetic Curie point (Hohenberg and Halperin (1977), as cited after the Introduction). We now look for something similar in the percolation field.

To describe such non-equilibrium phenomena, often also called 'transport properties', it is often not sufficient just to know everything about the static behaviour. An additional time-dependent property is needed, too; then one can try to express other transport properties through this time-dependent property and the static quantities. For percolation, we take the conductivity of random resistor networks as our basic transport property, and on its basis we will try to understand diffusion near the percolation threshold, as sketched briefly in the Introduction. Numerous other aspects of the kinetics of clusters were also discussed in the literature; we refer the interested reader to the conference proceedings edited by Family and Landau (1984).

5.1. Conductivity

Let us go back to the squares of Figure 1 which are randomly occupied or empty. We regard every occupied square as a piece of copper, whereas every empty square is regarded as insulating. An electric d.c. current can only flow between copper squares having one side in common, not between occupied squares touching at a corner only, or separated by even longer distances. How

Figure 23. Definition of the conductance of a random resistor network. All copper squares in the top-most row of the lattice are connected to a heavy copper bar (no loss of energy in the bar), and so are all squares in the bottom row. A battery then applies a unit voltage between these two bars. The resulting electrical current is called the conductance.

much current flows through the lattice if a unit voltage is applied between the top line and the bottom line of the lattice (see Figure 23)? We call this current due to a unit voltage the conductance of the sample.

We take the whole lattice to have a rectangular shape of $L \times N$ squares, with N being the length of the top (and bottom) row to which a uniform voltage is applied (see Figure 23). Both N and L are very large. Then the conductance is proportional to N and inversely proportional to L. In d dimensions it will still be inversely proportional to L, but it is now proportional to the cross-section, N^{d-1}, of the sample. Thus the conductance is proportional, in d dimensions, to N^{d-1}/L, and the factor of proportionality is called the *conductivity* Σ of the material, which is now independent of size and shape. For a square or cubic shape, one has $L = N$ and thus the conductivity Σ is L^{2-d} multiplied by the current produced by a unit voltage. (We set the distance between neighbouring points on the lattice equal to unity.)

It is obvious that for large lattices we have zero conductivity if no infinite network of neighbours is present, that is for $p < p_c$. When p is appreciably larger than p_c, nearly all copper squares have clustered together to form one infinite network, and the conductivity Σ as well as the fraction P of sites in the infinite network increase roughly linearly with concentration p. At $p = 1$ we have, of course, $P = 1$, and Σ reaches the conductivity of bulk copper. Since I am not an electrical engineer I am allowed to set this copper conductivity equal to unity, and thus $\Sigma(p=1) = P(p=1) = 1$. In other words, my copper squares carry a unit current if a unit voltage is applied to two opposing sides.

Because of this close relationship between conductivity $\Sigma(p)$ and mass $P(p)$ of the infinite network it would be nice if they were proportional (and because of our normalizations that would mean identical) over the whole range of p. Already the first experiment, by Last and Thouless (1971), has showed that this is not the case. They measured the current through sheets of graphite paper with randomly punched holes. Their results, shown schematically in Figure 24, indicate clearly that the two quantities Σ and P are not proportional. The conductivity versus concentration curve seems to end at the threshold with zero slope although a plot of P versus p has infinite slope there.

Why is this so? Dead ends contribute to the mass of the infinite network but not to the electrical current it carries. We learned earlier in Section 3.3 and Figure

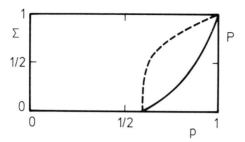

Figure 24. Conductivity Σ of conducting paper with holes randomly punched in it (solid line; from Last and Thouless 1971). The dashed line gives the percolation probability P. Obviously, the two quantities vanish with different exponents μ and β, though at the same threshold p_c.

17 that most of the mass of the infinite network at the threshold belongs to dead ends, not to the backbone. Thus most of the mass contained in P makes no contribution to the conductivity Σ, and therefore the critical exponent for Σ differs from the β for P. (Unfortunately, not even the backbone mass is proportional to the conductivity.)

We denote the conductivity exponent by μ:

$$\Sigma \propto (p - p_c)^{\mu} \tag{63}$$

for $p \to p_c$. Often this exponent is also called t but here we need t as the symbol for time (and earlier we used it for the perimeter.) Perhaps we should leave it as an exercise to the reader to derive a simple and exact scaling relation between this new exponent μ and the old ones like β, v, etc. for the people working in this field have not, at present, agreed on such a relationship. Various choices for μ were offered at various times in the last ten years for dimensionality d between 1 and 6:

$$\mu = 1 + \beta \qquad 1 + \beta' \qquad (d-1)v \qquad 1 + (d-2)v \qquad 1 + 2\beta \qquad (5d-6)v/4$$

$$\text{and} \qquad \mu = ((3d-4)v - \beta)/2 \tag{64}$$

with β' being for the backbone what β is for the full infinite network (Sahimi 1984).

The last three suggestions agree well with numerical data from $d=2$ to $d=6$, as partially listed in Table 2 of Section 2.7. (For high d, starting with $d=6$, the exponent μ takes on its Bethe lattice value $\mu = 3$.) However, none of them seems to work exactly in the whole range $1 < d < 6$. The Alexander–Orbach rule, Equation (64), will give a particularly simple result for the so-called fracton or spectral dimensionality to be mentioned later in diffusion theory and was thought to be correct when the first pages of this manuscript were written. This effect is quite common in modern research: hurry to publish before your result is shown to be wrong. Unfortunately, in this case I was not writing fast enough! A careful reader should inform himself about the state of the art at the time he is interested in this

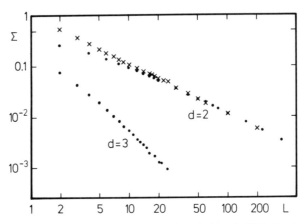

Figure 25. Variation of the conductivity Σ with system size L in three (solid line) and two dimensions (dashed lines), right at the percolation threshold. The two data sets for two dimensions refer to different geometries. From Derrida *et al.* (1984), Zabolitzky (1984), and Lobb and Frank (1984).

problem; perhaps the dust will have settled by then. It is, however, quite possible that the above relations are valid within certain intervals for the dimensionality d. Or perhaps there is no relation between μ and the other exponents, that is μ is a new 'time' exponent independent of the 'static' exponents β and ν. Except where explicitly stated we assume none of the above relations and work with μ as if it were an independent exponent.

Figure 25 shows some conductivities Σ as a function of system size L right at the percolation threshold. Finite-size scaling theory then tells us that

$$\Sigma \propto L^{-\mu/\nu}$$

The two-dimensional data give μ/ν between 0·97 and 0·98, in disagreement with the prediction 0·95 from Equation (64). One of these studies (Zabolitzky 1984) used a hundred hours on a Cyber 205 vector computer, the most ambitious percolation simulation made so far if measured by the number of operations. In three dimensions the figure tells us $\mu/\nu = 2\cdot2$, giving $\mu = 2\cdot0$. Series expansion (Adler) agree with these Monte Carlo studies.

How does one conduct such Monte Carlo studies? The easy part is to produce a lattice with conducting and insulating sites. The difficulty lies in calculating the conductance of that lattice. Kirchhoff's rules tell us that for every loop of conductors the sum of the voltages is zero, and that for every site the sum of the currents flowing into it is also zero. The resulting system of coupled linear equations for the voltages at every site can be solved by relaxation methods (Kirkpatrick 1973). The closer we are to the percolation threshold, the more iterations we need in this relaxation before we get the solution with the desired accuracy. While this 'critical slowing down' is of interest in itself, it also increases the computer time. That difficulty can be avoided if in two dimensions special transformations are used (see e.g. Lobb and Frank 1984) which allow the conductivity to be calculated by going just once through the lattice. In a similar spirit, but with an algorithm which is also useful for more than two dimensions, Derrida *et al.* (1984) calculated the conductivity of narrow strips (and bars) in their combination of Monte Carlo simulation with the 'striptease' renormalization of Section 4.2. We refer to these papers for more computational details.

Instead of working with resistors and insulators, one can also study a mixture of resistors and superconductors. In that problem each site has either zero resistivity (infinite conductivity) with probability p, or has a probability $1 - p$ of being a normal resistor. All quantum-mechanical aspects of real super-conducting materials are neglected, of course; we still deal only with geometry. The situation is formally similar to that of random resistor networks except that zero conductivity is replaced by zero resistivity. The conductivity of this network of superconductors and normal resistors is infinite whenever an infinite network of superconducting sites is formed, that is for $p > p_c$. It is finite below p_c; thus it is possible that this conductivity diverges as p_c is approached from below. In two dimensions it diverges with the same exponent μ with which the conductivity of resistor-insulator networks vanishes. In general, this nice duality rule is not valid. For example in three dimensions, where roughly $\mu = 2$, the superconducting exponent, often denoted as s, is only slightly larger than 0·7. Straley's rule says that this exponent equals $2 - \alpha - \mu$ whereas Kertész speculates it to be $v - \beta/2$. In two dimensions, where it equals μ, the first expression gives $\mu = v$, whereas the second one gives Equation (64); in three dimensions, Straley's rule seems to be more accurate. Again, the research front is quite fluid at present, and the problem may have been settled by the time you read this statement.

5.2. Ants in the labyrinth

In the Introduction we already mentioned the diffusion of a particle in a disordered network, a problem also dubbed the 'ant in the labyrinth'. Now we want to look at this problem from a more theoretical point of view to see whether we can apply what we have learned in the meantime. First let us repeat the definition of this kinetic process. At every time step, the diffusing particle, called the ant, selects randomly one of its nearest neighbour sites. If that site is occupied ('permitted', probability p), it moves there; if the neighbour is empty ('prohibited', probability $1 - p$), it stays where it is. That process is repeated again and again, and averaged over many different ants running through many different lattices. To avoid any X-rating of this book, ant–ant interactions are ignored. We are interested in the average distance R which the ants travel as function of the time t; t is the number of time steps (jump attempts) made by the ant.

We have already said in the Introduction that for p far above p_c one observes normal diffusion, $R^2 \propto t$, for large times, whereas for p far below p_c, R approaches a constant for long times. From what we have learned in the preceding chapters we can assume that for all concentrations above the threshold one type of behaviour (diffusion) will dominate asymptotically, whereas for all $p < p_c$ the other behaviour (finite asymptotic distance) will be valid. A third type, $R \propto t^k$, will govern the asymptotic distance right at the critical point. We will now study in greater detail the behaviour very close to p_c where these three different laws have to merge.

First let us look at ants diffusing within the finite clusters for p below p_c. An ant 'parachutes' from above to one occupied site, and then tries to move. If that site is completely isolated (i.e. if it is a cluster of mass $s = 1$), the ant cannot move at all, and the problem is solved. This is nearly always the case at very low p: $R(p \rightarrow 0) = 0$. For larger p, the probability is not negligible that the ant hits a triplet as shown in Figure 26 for one dimension; let us assume it hits in the centre. Initially, the probability of being at some site is therefore zero for the two ends, and unity for the centre. In the next step ($t = 1$), it selects randomly one of the two

●	●	●	●	●	●
0	1	0	0	1	0
1/2	0	1/2	1/2	0	1/2
1/4	1/2	1/4	0	1	0
3/8	1/4	3/8	1/2	0	1/2
5/16	3/8	5/16	0	1	0

Figure 26. A diffusion process has started in the centre of the cluster with $s = 3$. The numbers below each site give the probability that the ant is at that site, with the first line corresponding to $t = 0$, the second to $t = 1$, etc. The left part corresponds to the blind ant which selects any of its neighbours randomly at each step and thus may have to stay at its old site, whereas the right part corresponds to the 'myopic' ant which randomly selects one of its occupied neighbours and thus always moves there. $(d = 1.)$

neighbours both of which are permitted; thus the probability of being at one of the ends is 1/2 whereas the probability of being at the centre is zero. For the next step ($t = 2$), the ant, with equal probability, either tries to move outside the cluster (which is prohibited and therefore the ant has to stay where it was), or tries to move to the permitted centre site. The probability of remaining at one of the ends thus is 1/4, and that of being at the centre equals 1/2, at $t = 2$. In future steps, the differences between these probabilities will get smaller and smaller, as the numbers in Figure 26 indicate. The difference between the probability for the centre and that for the left end is $1, -1/2, 1/4, -1/8, 1/16, \ldots$ for $t = 0, 1, 2, 3, 4, \ldots$ and obviously goes to zero; the three probabilities each converge to 1/3.

Asymptotically every site is visited equally often by the ant. This fact has also been proven generally.

This result is by no means trivial, as the other example in Figure 26 shows. If at every step the ant selects randomly one of the occupied sites, then it always moves there. Now in our triplet the probabilities oscillate, and even their average values are not equal. From the point of view of percolation theory, this ant species is not a useful probe of our clusters since it prefers certain cluster sites to others. We therefore work only with the first species of ants which are more impartial and visit, in the long run, all sites of a finite cluster equally often. Ants on the right of Figure 26 are ignored, only the lefthand ants are relevant for us. (Experts also call our impartial and rather stupid ant 'blind' or 'drunken' whereas the second, more intelligent ant species is called 'myopic', 'half-drunken' or 'hasty' (Mitescu and Roussenq, as cited after the Introduction). It seems that the two species have the same critical exponents (Seifert and Suessenbach 1984).

This assumption of equal rights for the different sites of a cluster means that asymptotically the distance R for one finite cluster is the average distance between two cluster sites. This average in turn was our cluster radius R_s defined in Equation (45). Now we have to average over different cluster masses s. Initially an ant parachutes to each occupied site with the same probability. The probability of a site belonging to a cluster containing s sites is $n_s s$, and then the asymptotic distance is R_s. Usually one averages over the squared distances and then gets from the sum over all cluster sizes:

$$R^2(t=\infty, p<p_c)=\Sigma_s n_s s R_s^2$$

$$\propto (p_c-p)^{\beta-2v} \tag{65}$$

The critical exponent in this result is derived in the same way that we derived the exponent for Equation (31). We see that not all lengths are proportional to the correlation length ξ: the exponent is not simply $2v$. (Had we averaged over R_s instead of R_s^2, we would have got a critical exponent $\beta-v$; see again Mitescu and Roussenq for data.) In three dimensions, the exponent $\beta-2v$ is about -1.34. Monte Carlo determinations first gave much higher values (Mitescu and Roussenq), then agreed better with theory (Pandey *et al.* 1984, Seifert and Suessenbach 1984). Let us thus hope that we understand the behaviour below the threshold; what about above and at p_c?

For $p=1$ we showed already in the Introduction that $R^2=t$ exactly for all t, not only asymptotically. Generally one calls any law $R^2 \propto t$ a diffusion law, and we denote the proportionality constant as the diffusivity D:

$$R^2 = Dt$$

for long times t. Thus for $p=1$ we have diffusion with $D=1$, even for short times. (The usual definition of D differs from ours by a trivial d-dependent factor, which we ignore to have $D(p=1)=1$.) For p slightly below unity but still far above p_c we have some holes in the lattice which slow down the ant but still let it diffuse everywhere. Thus the diffusion law should still be valid, only with a reduced diffusivity $D = D(p)$. For p below p_c diffusion is impossible since now $R^2(t)$ is bounded by Equation (65). Thus for very large times R cannot increase as $t^{1/2}$, and the diffusivity is zero. Thus $D(p)$ goes to zero if p approaches p_c from above.

Fortunately, no new critical exponent is needed to describe how D vanishes near the percolation threshold. De Gennes (1976) in the French review mentioned after the Introduction, pointed out that the diffusivity D is proportional to the conductivity Σ of random resistor networks. Since we both normalized them to unity at $p=1$ and since de Gennes' proportionality is not restricted to critical phenomena, we simply have

$$D = \Sigma \quad \text{or} \quad R^2 = \Sigma t (t \to \infty) \tag{66a}$$

This equation is simply a manifestation of Albert Einstein's result from the beginning of this century, that in statistical physics the diffusivity is proportional to the mobility. The mobility, on the other hand, is the ratio of the velocity to the applied force. For the electrons in the copper parts of a random resistor network, the applied force is proportional to the electric field, that is to the voltage. The average velocity of the electrons is proportional to the electrical current they produce. Thus their mobility is proportional to the ratio of current to voltage, that is to the conductivity of the network. Thus Equation (66a) is basically due to Einstein and therefore needs no further proof. (Computers checked it, anyway.) Close to the percolation threshold we recover our conductivity exponent:

$$D \propto (p - p_c)^\mu \tag{66b}$$

How can we combine the two so seemingly different results (Equations (65), (66)) into one consistent theory? Having studied Equations (33), (52), (54) the reader will have no difficulty in recognizing that scaling can again be applied to the distance R depending on two variables $1/t$ and $p - p_c$ both going to zero. The general statements discussed after Equation (52) also apply here. (A warning: The analogy with our earlier results is closest if we regard the time t as function of the distance R and replace the system length L by the ant distance R in Equation (54). But that last step would be incorrect since R according to Equation (65) does not scale as ξ.) With two suitable exponents x and k and a scaling function $r(z)$ we assume

$$R = t^k r[(p - p_c)t^x] \tag{67}$$

For p above p_c and sufficiently long times and distances we must recover the diffusion law of Equation (66); thus for large positive arguments z the scaling function $r(z)$ is proportional to $z^{\mu/2}$ in order to be consistent with Equation (66b). Then

$$R \propto t^k (p - p_c)^{\mu/2} t^{\mu x/2} \propto t^{k + \mu x/2} D^{1/2}$$

We will also need $R \propto t^{1/2}$ in this regime, which requires $k = (1 - \mu x)/2$. (Our R is the root mean square distance, that is the square root of the averaged R^2.)

On the other hand, for p below p_c we must recover the result given by Equation (65) that R varies as $(p_c - p)^{-v + \beta/2}$ independent of t for sufficiently long times. The scaling function $r(z)$ for $z \to -\infty$ thus must vary as $(-z)^{-k/x}$ in order that t cancels out: $R \propto (p_c - p)^{-k/x}$. Equation (65) now requires this exponent k/x to equal $v - \beta/2$, or $k = (v - \beta/2)x$. Equating these two expressions for k we get $1 - \mu x = (2v - \beta)x$, or

$$x = 1/(2v + \mu - \beta) \tag{68a}$$

from which

$$k = (v - \beta/2)/(2v + \mu - \beta) \tag{68b}$$

follows.

We have thus derived the two exponents x and k entering the scaling assumption (67). If we simulate the ant right at the critical point, we get an 'anomalous diffusion' exponent k smaller than the usual $1/2$ from Equation (67):

$$R \propto t^k \propto t^{(v - \beta/2)/(2v + \mu - \beta)} \tag{69}$$

for long times t and $p = p_c$. This anomalous exponent k is about 0·33 in two dimensions (it would be $1/3$ exactly if the Alexander–Orbach rule of Equation (64) were exact) and close to 0·2 in three dimensions. Numerous numerical tests have confirmed these predictions with reasonable accuracy. Figure 27 shows three-dimensional data whereas Figure 5 in the Introduction has already given two-dimensional results.

We see from the argument z of the scaling function $r(z)$ in Equation (67) that there is a characteristic time in our relation between R and t. For times much smaller than $|p - p_c|^{-1/x}$ but much larger than unity one has anomalous diffusion, Equation (69), whereas for much longer times one either observes normal diffusion, Equation (66), or a constant distance, Equation (65). The characteristic time $\propto |p - p_c|^{-1/x} = |p - p_c|^{\beta - 2v - \mu} \propto |p - p_c|^\beta \xi^2/D$ separates the regime of anomalous diffusion from the more usual behaviour.

You may wonder why this characteristic time is not simply proportional to ξ^2/D, the time needed for an ant to diffuse over a distance of the order of the correlation length ξ. Where does the additional factor $|p - p_c|^\beta$ come from? The reason is the same as that for Equation (65). It comes from averaging over all cluster sizes. Percolation properties are usually derived as being the sum of the

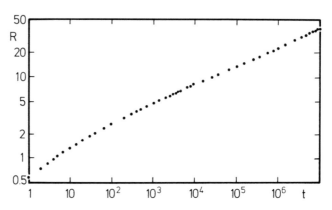

Figure 27. Root mean square distance R travelled by a diffusing particle ('ant') in t step attempts on a simple cubic lattice at its percolation threshold. Pandey *et al.* (1984) determined $k = 0.2 \pm 0.01$ from these results.

contributions from single clusters. Thus a more microscopic scaling theory (Gefen, Aharony, Alexander) would start with a generalization of Equation (67) to the distance R travelled by an ant during the time t in a cluster containing s sites: $R = R(s, t, p)$ instead of the above $R(t, p)$. Summation over all s, as in Equation (65), then has to give Equation (67) and the other results above.

If we now look for p slightly above p_c at very large clusters only, with s much larger than $|p - p_c|^{-1/\sigma}$, and with cluster radii much larger than both the correlation length and the distance travelled by the ant, then the ant will not feel the cluster boundaries. Thus it will behave as if it were on the infinite cluster. There it diffuses as usual: $R^2 = D't$. This new diffusivity D', however, is not the same as our diffusivity D defined above for the case where the ant starts running on an arbitrary occupied site, not necessarily on the infinite cluster. If it starts anywhere on an occupied site, then with a probability P/p it starts on the infinite cluster, and with the probability $1 - P/p$ it starts on a finite cluster. For the strength P of the infinite cluster is the probability that an arbitrary lattice site, occupied or empty, belongs to the infinite cluster. Only an ant on the infinite cluster can contribute to a distance increasing with time ($R^2 = D't$); the other ants add only a finite amount to the distance. Thus, if we average over all occupied sites as starting points of the random walk we get for sufficiently long times:

$$Dt = R^2 = (P/p)D't$$

The diffusivity D in the whole lattice, and the diffusivity D' in the infinite network or very large cluster are thus related by

$$\frac{D}{D'} = \frac{P}{p} \qquad (70)$$

an equality not restricted to the region very close to the percolation threshold.

Now we see that our characteristic time $|p - p_c|^{-1/x}$ discussed above is nothing but the time ξ^2/D' the ant needs to travel the distance ξ if it diffuses, with diffusivity D, on an infinite or very large cluster. For close to the threshold we have $D' \propto D/(p - p_c)^\beta$, and thus we have explained the unexpected factor $|p - p_c|^\beta$ in our characteristic time. This factor is not a violation of scaling but merely indicates that different types of averages may have different critical exponents.

For a theory of diffusion on finite clusters with radius R_s it is not sufficient to assume

$$R = R_s r[(p - p_c)t^x]$$

in analogy with Equation (67). Instead the scaling function r depends also on the scaling variable $(p - p_c)s^\sigma$. We do not go into these details (Gefen *et al.* (1983), Wilke *et al.* (1984), Pandey *et al.* (1984), Kertész and Metzger (1983)) since we want to restrict this book to scaling functions of a single variable only.

Monte Carlo studies have been made both with ants running everywhere where permitted (e.g. Pandey *et al.* (1984)), or being restricted to the infinite cluster (e.g. Havlin and Ben Avraham (1983)). The latter case can be treated with theories analogous to those above for the unrestricted case, only with D replaced by D'. Then some exponents μ are replaced by $\mu - \beta$, and right at the critical point we can write the analogue of Equation (69) as

$$R \propto t^{k'} \qquad k' = \frac{v}{(2v + \mu - \beta)} \tag{71}$$

for $p = p_c$ and $t \to \infty$. This anomalous diffusion coefficient is about 0·35 in two and about 0·27 in three dimensions and is roughly confirmed by Monte Carlo studies on the incipient infinite network.

If the reader studies the literature on this subject he should always be careful to note whether the paper deals with diffusion anywhere, or only on the infinite cluster. Also the notation for various exponents is different in different papers. Some authors call the reciprocal of k or k' the fractal dimension of the walk; it differs, of course, from the fractal dimension of the cluster on which the walk takes place.

The scaling assumptions like Equation (67) have counterparts in thermal critical phenomena (Hohenberg and Halperin (1977), cited after Chapter 1). There one often studies systems by light or neutron scattering, as a function of frequency ω and wavevector Q, which are analogous to our $1/t$ and $1/R$, respectively. Then the analogue of Equation (67) is

$$Q = \omega^k q[(T - T_c)\omega^{-x}]$$

The characteristic time diverges at the critical point as $|T - T_c|^{-1/x}$; often that exponent is denoted by Δ. Right at the critical point $T = T_c$ we get

$$Q \propto \omega^k \quad \text{or} \quad \omega \propto Q^{1/k}$$

Often the exponent $1/k$ is called z in these dynamical scaling theories. Theories of this kind describe, for example, the transition from a normal spin wave spectrum, $\omega \propto k^2$, in an isotropic ferromagnets to anomalous diffusion, $\omega \propto Q^z$, at the critical point, and to normal spin diffusion in the paramagnetic phase. The analogy is even closer in a dilute ferromagnet at low temperatures (Harris and Stinchcombe, 1983). Then the proportionality constant in the spin wave spectrum is nothing but our diffusivity in percolation theory.

5.3. Kinetic clusters within static clusters

What does the set of sites visited by an ant running t steps in the labyrinth look like? If no site could be visited twice, as in self-avoiding walks, this set of visited sites would contain $t + 1$ occupied sites; but our walks are not self-avoiding. On a completely occupied lattice ($p = 1$), the number V of different visited sites increases asymptotically as t^ψ in d dimensions, with $\psi = d/2$ or $\psi = 1$, whichever is smaller (Rammal and Toulouse 1983). On a finite cluster with s sites one has asymptotically $V = s$. A general scaling assumption for $V = V(p, s, t)$ again involves a scaling function of two variables. Let us thus restrict ourselves to $p = p_c$ and ask for the number $V(s, t)$ of different sites visited by an ant diffusing t steps on a large but finite cluster containing s sites. Our scaling assumption for long times then is

$$V = t^\psi v(t/s^y) \tag{72a}$$

at $p = p_c$, with a suitable scaling function v and two exponents y and ψ to be determined now.

For very long times, V must equal s independent of t, and thus the scaling function $v(z)$ equals $z^{-1/y}$ for large z. This decay in turn requires $\psi = 1/y$ in order that t cancels out:

$$V = t^\psi v(t/s^{1\psi}) \tag{72b}$$

On the other hand, t should scale as the characteristic time $|p - p_c|^{\beta - 2\nu - \mu} \propto \xi^2/D'$ of the preceding Section 5.2. From the scaling of cluster numbers, Equation (33), we know that $|p - p_c|$ scales as $s^{-\sigma}$. Therefore $s^{1\psi}$ in Equation (72b) should equal $s^{(2\nu + \mu - \beta)\sigma}$, or

$$\psi = \frac{1}{\sigma(2\nu + \mu - \beta)} = \frac{k'}{\rho} \tag{73}$$

with $1/D = \rho = \sigma v$ from Equation (50) and with k' from Equation (71). (2ψ is also called the fracton or spectral dimension.) If the Alexander–Orbach rule of Equation (64) were correct, this exponent ψ would be exactly 2/3, independent of d. Then we would have

$$k' = \frac{2\rho}{3} \tag{74}$$

which allows predictions to be made for all fractals of dimension $1/\rho$ even if μ, v and β are undefined (Wilke *et al.* 1984). Such a 'super-universal' critical exponent would indeed be very nice. Figure 28 shows Monte Carlo data on the incipient infinite cluster in 2 to 6 dimensions. If $\psi = 2/3$, then the ratio $V/t^{2/3}$ plotted there

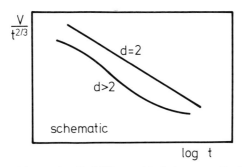

Figure 28. Variation of the number V of different visited sites, normalized by the theoretical value $t^{2/3}$, with the time t of the diffusion process on the incipient infinite cluster in a d-dimensional lattice. The number on each curve gives d; from Rammal *et al.* (1984).

versus time would approach a constant $v(0)$ for large times, Equation (72b). We see that the data, except perhaps for $d = 2$, are roughly consistent with this approach to a constant. Actually, it is this simple 2/3 law which also enters into other quantities (Rammal and Toulouse 1983), through which Equation (64) was first found. Unfortunately, the more direct determination of μ in two dimensions indicates slight deviations from this nice super-universality (e.g., Zabolitzky 1984). Perhaps at the time you read these remarks the situation will be clearer.

The set of all V sites visited by the ant constitutes another cluster, within the fixed cluster onto which the ant parachuted initially. Thus we have here another 'dynamic' cluster imbedded by the diffusing ant into the pre-existing static cluster. Research is just beginning into the geometrical properties of these clusters within clusters (see e.g. Stanley *et al.* 1984).

Many different ant species have been investigated. We have already mentioned the difference between the democratic blind ant we were working with here and the elitist myopic ant which does not visit all sites with equal probability. In addition, ants can have inertia and thus move straight ahead with an enhanced probability. Or they tend to drift in one certain direction, like electrons in an electric field, or prefer circles like electrons in a magnetic field. Finally they may be attracted or repelled by occupied sites, or by each other. In some cases the asymptotic critical behaviour changes, in others not. 'Termites' is the name for those insects which move very fast on a cluster, and dig tunnels very slowly through the empty regions between clusters; computers are trying to train termites (Adler *et al.* 1985, Bunde *et al.* 1985) with a diffusivity proportional to the conductivity of superconductor-resistor networks. Finally, 'parasites' (Wilke *et al.* 1984) are (what else could they be), ants on large lattice animals. Before your home gets completely infected with all these insects, I will stop putting more of them in front of you. However it seems likely that in the literature published between the writing and the reading of this book, more kinetic processes in disordered media will be studied (Family and Landau 1984).

Further reading

A general recent review on diffusion research was given by K. Kehr and K. Binder in Chapter 6 of *Applications of the Monte Carlo Method in Statistical Physics* (Heidelberg: Springer-Verlag, 1984), edited by K. Binder.

The following literature was cited in the text:

Adler, J., *J. Phys. A*, **18**, 307, (1985).

Adler, J., Aharony, A. and Stauffer, D., *J. Phys. A*, **18**, L 129, (1985).

Alexander, S. and Orbach, R.,, *J. Physique (Paris) Lettres*, **43**, L 625, (1982).

Bunde, A., Coniglio, A., Hong, D.C. and Stanley, H.E., *J. Phys. A*, **18**, L 137, (1985).

Derrida, B., Zabolitzky, J.G., Vannimenus, J. and Stauffer, D., *J. Statist. Physics*, **36**, 31, (1984).

Family, F. and Landau, D.P. (eds), International Conference on *Kinetics of Aggregation and Gelation* at Athens, USA, April 1984, (Amsterdam: North Holland Publ. Company, 1984).

Gefen, Y., Aharony, A. and Alexander, S., *Phys. Rev. Letters*, **50**, 77, (1983).

Harris, C.K. and Stinchcombe, R.B., *Phys. Rev. Letters*, **53**, 1399, (1983).

Havlin, S. and Ben Avraham, D., *J. Phys. A*, **16**, L 483, 1983).

Kertész, J., *J. Phys. A*, **16**, L 471, (1983).

Kertész, J. and Metzger, J., *J. Phys. A*, **16**, L 735, (1983).

Kirkpatrick, S., *Rev. Mod. Phys.*, **45**, 574, (1973).

Last, B.J. and Thouless, D.J., *Phys. Rev. Letters*, **27**, 1719, (1971).

Lobb, C.J. and Frank, D.J., *Phys. Rev. B*, **30**, 4090, (1984).

Pandey, R.B., Stauffer, D., Margolina, A. and Zabolitzky, J.G., *J. Stat. Phys.*, **34**, 427, (1984).

Rammal, R. and Toulouse, G., *J. Physique (Paris) Lettres*, **44**, L 13, (1983).

Rammal, R., Angles d'Auriac, J.C. and Benoit, A., *Phys. Rev. B*, **30**, 4087, (1984).

Sahimi, M., *J. Phys. A*, **17**, L 601, (1984).

Seifert, E. and Suessenbach, M., *J. Phys. A*, **17**, L 703, (1984).

Stanley, H.E., Majid, I., Margolina, A. and Bunde, A., *Phys. Rev. Letters*, **53**, 1706, (1984).

Straley, J., *J. Phys. A*, **13**, 819, (1980).

Wilke, S., Gefen, Y., Ilkovic, V., Aharony, A. and Stauffer, D., *J. Phys. A*, **17**, 647, (1984).

Zabolitzky, J.G., *Phys. Rev. B*, **30**, 4077, (1984).

CHAPTER 6

summary

Once your nightmares are over about termites digging tunnels through your brain, corruption ruling in renormalization country, or 1·56 dimensional animals burning in forest fires, you may wish to reflect on whether you have learned anything from this book.

My first aim was to show that percolation is an active field of research. Many of the results prresented here were not yet known at the time most of the prospective readers of this book were born. In general, the knowledge was formed in the order in which the chapters were printed. Thus the kinetic aspect contains, as its reading list suggests, the most recent ideas, in particular in Section 5.3. A theorem by Dyson tells us that a publication coming out at time t and taking into account research up to time $t-t_0$ will be outdated at time $t+t_0$. Indeed even while I was writing this book I had to modify earlier parts due to new results coming out. If you prepare any seminar talk about some part of this book, you should therefore try to find out what has happened in research more recently, for example by consulting the Science Citation Index after locating a relevant publication from the reading list in this book. In short, theoretical physics in general and percolation theory in particular is a human enterprise and not the fixed body of knowledge which it often appears to be when presented in formal courses within your curriculum.

In addition I hope you have learned what a phase transition is. Admittedly, percolation has a somewhat unusual phase transition since no temperature is involved. But nevertheless, many functions or their derivatives, as a function of a continuously varying parameter p, diverge or vanish at one sharply defined point, the percolation threshold $p = p_c$. In percolation theory these functions are purely geometric properties; at most other phase transitions one deals with thermal properties like specific heats as function of temperature etc. Nevertheless, both cases are similar in that the important functions, or their derivatives, are not continuous at the critical point.

The similarity between thermal phase transitions and the percolation threshold becomes even clearer when we look at the scaling laws governing the leading asymptotic behaviour very close to the critical point. Many functions

$G(x, y)$ depend on two variables x and y which both vanish right at the critical point. Then, in Equations (33), (52), (54), (67), (72) we have seen five different examples of the same principle:

$$G(x, y) = x^{-A}g(y/x^B)$$

is the scaling assumption for small x and small y. For example, y may correspond to the distance from the critical point, and $1/x$ to the cluster size. Scaling theory does not predict the value of the critical exponents A and B or the precise form of the scaling function g; but with this scaling assumption we can calculate the critical exponents of many other quantities and relate them to A and B. In general, for simple problems two independent exponents like A and B here are sufficient to determine the other exponents. No clear answer could be given, for percolation as well as for thermal phase transitions, as to whether or not the additional exponent entering into dynamical scaling is related to the static exponents. For thermal phase transitions that depends on the system studied, for percolation it is at present controversial whether the conductivity exponent is related to the static critical exponents. To calculate the numerical value of any of these exponents we have to go beyond scaling theory since then relations between exponents are not enough. Renormalization group techniques have been developed into a powerful tool to estimate exponents numerically. In a few cases, mostly in two dimensions, we (believe to) know even the exact values.

Right at the critical point, one of the arguments in the function $G(x, y)$ above may be zero; for example $y = 0$ if $y = p - p_c$ or $y = T - T_c$. Then most quantities vary asymptotically with a simple power law, like $G \propto x^{-A}$ for small x. If x happens to be a length one may, under certain conditions of self-similarity, call $-A$ the fractal dimension for the quantity G, particularly if G can be identified with the mass on an object. In the largest cluster right at the percolation threshold of an $L \times L \times L$ simple cubic lattice, the number of sites increases asymptotically as $L^{2.5}$, which makes it a 2·5 dimensional fractal. We see that in this sense the concept of fractals is contained in the concept of scaling near phase transitions; but the fractal concept can also be applied to power laws where no phase transition occurs, like the lattice animals and their radii discussed after Equation (51).

There is no need to understand thermal physics before one studies percolation. One does not need to know classical or quantum mechanics, or statistical physics, to understand percolation. Only geometry and probability, and for conductivities some elementary concepts of electricity, were required in this book. On the other hand, the knowledge from our simple percolation can be helpful in understanding better the behaviour of the more complicated thermal phase transitions.

Finally I hope you have learned that percolation is a good introduction to the field of computer simulations. The largest systems ever simulated (10^{10} sites) seem to belong to percolation theory. Small systems can be simulated on simple computers and produce nice pictures like Figure 2. Percolation theory therefore can and has been used as a way to learn computer usage. For the expert our appendix will give somewhat more information, since the basic difficulty of computer experimentation is the interpretation of data, not their production.

We call these computer simulations 'experiments' since in this book we avoided contact with real experiments in the laboratory or in Nature. Such real experiments, of course, have some difficulties. Usually the lattice is not completely periodic but has some defects, some impurities cannot be avoided, the distribution of sites is not completely random, etc. Therefore this book tried to introduce the reader to percolation theory, not to percolation experiment. (We saw, however, that computer experiments have difficulties too: The systems are quite small, and for very large systems the random numbers might not be random enough.) These deviations make it in general quite difficult to get an exact correspondence between the percolation model and some real material. However, the principle of universality suggests that the critical exponents found in percolation theory agree with the critical exponents which careful laboratory experiments should give. If nearly all three-dimensional percolation models have the same critical exponents, then these exponents should not depend on such minor problems as lattice defects. This universality is one of the reasons why critical exponents were emphasized so much in this book. Indeed, certain metal-insulator films turned out to mimic excellently the behaviour found in computer simulations of percolation models, as reviewed by Deutscher, Kapitulnik, and Rappaport in the book on *Percolation Structures and Processes* mentioned after the Introduction. In the same book, Zallen gives a beautiful overview of the many different applications which one might find for percolation.

Let us take just one of these examples, the gelation of polymers. This phase transition was described as percolation on the Bethe lattice by Flory and Stockmayer during World War II, though they used different words. Therefore historically it constitutes the first application of percolation concepts. Even today, however, we do not know reliably whether the critical exponents of gelation are the same as those of percolation, as reviewed by Jouhier *et al.* in the book mentioned above, or by Stauffer, Coniglio and Adam (1982) in the article mentioned after the Introduction. Obviously the development of theoretical concepts and their tests by computer simulation are simpler than precision measurements on materials occurring in nature, and this book emphasized, in this and in other aspects, the simple problems, not the complicated ones. I tried to tell you the truth and nothing but the truth; but I never promised you the whole truth because nobody knows it.

APPENDIX

numerical techniques

In this appendix we discuss how one can estimate asymptotic quantities from exact 'series' data or from Monte Carlo simulations, and we explain how a computer can count clusters in very large lattices. Estimates of asymptotic properties are usually quite accurate. Estimates of systematic errors, on the other hand, are similar to weather prediction: one can never be sure of being right if one has predicted the behaviour at infinity for a quantity known only for finite intervals. Beginners in particular tend to underestimate the size of the systematic errors involved in their extrapolations.

A.1. Analysis of exact data

We have mentioned in Section 2.7 two methods of series analysis: Ratio and Padé approximation. To apply these methods, we need to know from exact enumeration the coefficients of a power series in the concentration p or in some other suitable variable. Such methods are also important for thermal critical phenomena. In percolation, we often have data on cluster numbers n_s, cluster radii R_s, numbers of lattice animals g_s, or other similar quantities which we want to analyse directly and which are not coefficients of a series.

In this case we may simply introduce the so-called generating function

$$G(\lambda) = \Sigma_s g_s \lambda^s \qquad (75)$$

where g_s stands for the quantity we are interested in. If asymptotically for large s the numbers g_s vary as

$$g_s \propto s^{-\theta}(const)^s \qquad (76a)$$

as is the case with many of these quantities, then we can evaluate G for λ slightly below $1/const$ by replacing the sum in Equation (75) with an integral.

Straightforward integration, similar to the evaluation of moments of the cluster size distribution in Chapter 2, then gives

$$G \propto \varepsilon^{\theta - 1} \tag{76b}$$

for small $\varepsilon = -\ln(\lambda \, const) \propto const^{-1} - \lambda$. This generating function and its critical exponent $\theta - 1$ can then be analysed with Padé approximations. (If $\theta > 1$ it may be practical to look instead at the generating function of $s^k g_s$ with k chosen to be so large that the generating function diverges at $\lambda = 1/const$. Equivalently one may look at derivatives of $G(\lambda)$.)

The ratio analysis of the generating function $G(\lambda)$ deals more directly with the 'expansion coefficients' g_s, and then we may even forget the concept of the generating function completely. What we described as the ratio method in Section 2.7 simply corresponds to looking at g_{s+1}/g_s, which equals $(1 + 1/s)^{-\theta} const$, if Equation (76a) can be applied. Plotting these ratios versus $1/s$ should give straight lines (for large s), with intercept $= const$ and slope $= -\theta \, const$; for we may use $(1 + 1/s)^{-\theta} = 1 - \theta/s$ for large s.

Such a plot gives a rough estimate of the exponent as well as the constant and also shows us from which size of s onwards we can utilize our data to get good extrapolations to infinite s. For higher precision one should use more subtle methods, provided one has many data for large enough s. For example, one may calculate the quantity

$$\frac{g_{s-1} g_{s+1}}{g_s^2}$$

If Equation (76b) is valid for large enough s, this quantity equals $(1 - 1/s^2)^{-\theta}$; thus from every triplet of consecutive numbers g_s one can calculate an estimate $\theta(s)$ for the exponent θ. There is no need to know the constant in Equation (76) since that has cancelled out. Now one may look at the variation of these effective exponents $\theta(s)$ with s.

θ can be plotted against $s^{-\Omega}$ with a 'correction-to-scaling' exponent Ω determined such that the data follow a smooth curve with a finite slope for $1/s \rightarrow 0$, and a monotonic behaviour of slope and curvature for large s. Such a plot corresponds to the assumption that the RHS of Equation (76a) is multiplied by a factor $(1 + C/s^\Omega)$ for large s; in general such assumptions are expected to be true.

Alternatively $\theta(s)$ is plotted against $1/s$, a straight line is drawn through each pair of consecutive points, and the intercept is determined. These intercepts will be another set of estimates $\theta(s)$ for the wanted exponent. You may repeat this game by plotting these intercepts against $1/s$, etc. etc. etc., until your data oscillate so strongly that they are useless. In this way you get some feeling for the errors involved in your estimate.

In general one should try several methods; while one of them may turn out to give the best fit to the data, the differences between this and other methods are an estimate of the errors involved. In the above tricks, 'plotting' does not necessarily mean drawing points on a piece of paper. One can also calculate the intercepts by computer or hand calculator. However, in general one needs many digits in the raw data and the ratios and other quantities derived from them. Thus on 32-bit computers I recommend double precision like Real*8 in IBM Fortran. (Computer time for the analysis is negligible even then compared with the computer time needed to produce the raw data g_s.) For the same reason, Monte Carlo data in general do not allow such ratio plots and similar tricks since their accuracy is not sufficient. The advantage of the exact results for small s is that at least the single data points have no error bars.

A.2. Analysis of Monte Carlo data

Monte Carlo data, in contrast to the exact 'series' data mentioned above, have finite statistical errors. Thus except for very high quality data, it is not recommended to fit straight lines through two consecutive data points. In general the slope and intercept of these lines will fluctuate too strongly to be useful. In fact, many Monte Carlo data are not precise enough to estimate three parameters simultaneously with reasonable accuracy, as in Equation (76) (constant, exponent θ, and factor of proportionality). Fortunately, for percolation, the critical point p_c can be determined well by the iteration method of Section 2.7, independent of the data for any divergent quantity or generating function. Thus for medium-quality data we can take the percolation threshold from that method, or from the literature; then only two parameters are needed to describe the leading asymptotic behaviour of many quantities.

For example, we may look at the mean cluster size $S(p) \propto |p - p_c|^{-\gamma}$ near the threshold, or at the cluster size distribution $n_s(p_c) \propto s^{-\tau}$ at the threshold for large s. Then with medium quality data one plots S or n_s double-logarithmically versus $|p - p_c|$ or s, respectively. Fitting a straight line through the data points gives an estimate for the exponent γ or τ from the slope, whereas the intercept gives the factor of proportionality. If $p - p_c$ or s vary by at least an order of magnitude, without taking into account data for away from p_c or small clusters, respectively, then the exponent estimates may be correct within about ten per cent.

This fit of a straight line may be done visually on log-log paper or it may be done electronically. Many hand calculators have built in programs to determine slope and intercept automatically. If you want to program this fit for yourself you will find the formulae in suitable manuals or textbooks (least squares fit). They are also derived easily by assuming that (taking $y_i = \ln(S)$ as function of $x_i = \ln|p - p_c|$) the sum of the squared deviations, $\Sigma_i(y_i - ax_i - b)^2$, from the straight-line fit $y = ax + b$ is as small as possible. Differentiating this sum with respect to a and b, and setting the results equal to zero, determines the slope a (exponent) and the prefactor b (proportionality factor).

Each single datum point has a statistical error. A rough estimate of this statistical error can be made by repeating the experiment with another set of random numbers. The difference then is of the order of the statistical error. A more reliable estimate is to make not just two but N independent simulations. We denote the average of a quantity y over these N simulations by $\langle y \rangle$, at one fixed parameter x. Then the probable statistical error (which is not the standard deviation) for y is

$$\Delta y = \left[\frac{\langle y^2 \rangle - \langle y \rangle^2}{N-1} \right]^{1/2} \tag{77}$$

If we are interested in only one value for x we can write $\langle y \rangle \pm \Delta y$ as our estimate and error bar for the quantity y. (The denominator $N-1$ is needed to account for the fact that your error is smaller the harder you have worked, that is the larger N is. On the other hand, with only one measurement, i.e. for $N = 1$, one has $\langle y^2 \rangle = \langle y \rangle^2$ even though the error is not zero. Thus it would not be legitimate to divide by N only. In any case, N should be large for precision measurements, and then the difference between N and $N-1$ is unimportant.)

Unfortunately, to estimate asymptotic properties these statistical errors are not enough. Usually we do not want to know what the cluster numbers are for $s = 1000$, but how they behave for $s \to \infty$, knowing them only for, say, $10 < s < 100$. Even if the statistical errors are exactly zero (as they are in series data) we cannot extrapolate to infinity with zero error. The additional deviations due to these necessary extrapolations are called systematic errors. Textbooks often contain formulae showing how to estimate the statistical errors for the slope and intercept of a straight-line fit from the statistical errors Δy of the single datum points. But these formulae are misleading for our applications since they ignore the systematic error. From only two exact points, say n_s for $s = 1$ and $s = 2$, these formulae would predict zero error for exponent and proportionality factor in $n_s \propto s^{-\tau}$. But we know that we cannot predict the behaviour for very large s with zero error if we know it only for $s = 1$ and $s = 2$.

How can we get better estimates for the quantities of interest and their error bars? First we need high precision for each point. Then, if we plot, say, n_s versus s double logarithmically, we will see curvature in our data even for rather large s. This curvature tells us that a simple power law is not sufficient to describe the data. Thus instead let us try to work with

$$n_s(p_c) \propto s^{-\tau}(1 + const/s^\Omega + \cdots) \tag{77}$$

with one correction-to-scaling term proportional to $s^{-\Omega}$. Analogous assumptions can be made for, say, the mean cluster size $S(p)$ as function of $p - p_c$(see review of Adler *et al.* in the book on *Percolation Structures and Processes* mentioned after the Introduction.) Of course, there will be more than just this one correction term to the leading $s^{-\tau}$ variation, but in general present series or Monte Carlo data are not accurate enough to determine a second correction term reliably if nothing else is known about it.

How do we estimate the parameters in Equation (77)? If you do not have a computer program at your disposal, fitting expressions like Equation (77) with many free parameters on a given set of data, then you should do this by yourself. That seems less professional but has the advantage that you know better what is happening in the analysis. Also, since different methods should be tried you get a better impression of systematic errors etc. by yourself than if you let a computer handle the whole analysis. In the above example (see article of Margolina *et al.* mentioned after Chapter 2) you may simplify the work by taking into account scaling laws. The exponent τ may be found more accurately from $\tau = 2 + \beta/(\beta + \gamma)$ than from the cluster numbers directly. Assuming this relation to be valid, you may then plot $s^{\tau} n_s$ versus $s^{-\Omega}$ for a certain trial value of Ω. If the data follow a smooth curve with monotonic behaviour and a finite slope at the intercept, then the tangent on this curve is a straight line, the intercept of which gives the proportionality factor in $n_s \propto s^{-\tau}$, and the slope of which is related to the constant appearing in the correction term of Equation (77). Then Ω is varied until the most satisfactory straight-line fit is found. The error bars in Ω may be estimated by calculating the sum of the squared deviations in this best fit (where the sum should be close to a minimum as function of Ω), and then by changing Ω away from the best fit until this sum is twice as large as in the minimum. The resulting spread in Ω is a plausible error bar. (If you make the fits on your hand calculator it may give you the correlation coefficient, which is very close to unity for good fits. Instead of calculating the sum of the squared deviations you may also work with the difference between unity and this correlation coefficient. If you have data of even quality these various methods will give about the same results.)

Alternatively you may set Ω equal to unity (a plausible but not reliable value) and vary τ until you get a nice straight-line fit. If neither τ nor Ω are known (or whatever corresponds to these quantities in your problem), then you may make the above analysis for one value of τ and determine the Ω with the nicest fit. Repeating that procedure for different τ you may get a curve $\Omega(\tau)$ corresponding to the valley in the surface representing the sum of the squared deviations as function of τ and Ω. The deepest point of that valley is then your best estimate. To do this analysis efficiently you should have at least a programmable hand calculator so that you can store the raw data once and for all in its registers. But still the most complex single procedure is the usual straight line fit.

Whichever method you use, you should also vary the set of data points on which you fit the straight lines. Usually, your data for small clusters (or for p far away from p_c) are more accurate than those for large s (or p near p_c). Unfortunately we need mainly data for large clusters. Omitting the largest and/or the smallest cluster size will slightly change your estimates and give you a better impression of the systematic errors in your analysis.

If you have data varying over a large range of parameters, like the distances in Figure 27 as a function of time up to 10^7, you may also go back to the good old log-log plot, determine the slope of the curve as a function of one of the variables, and extrapolate this slope to the asymptotic regime by plotting this slope versus some suitable power (exponent Ω) of the selected variable.

Generally, in a careful analysis you should try all methods which seem suitable. The spread of the results may give you a better impression of the error bars for your estimate than any one single method.

Unfortunately, even that is not enough. If you have found a suitable fit for, say, $n_s(p_c)$ versus s, you may still get something wrong since you had to employ finite systems. Sometimes (see Section 4.1) the effects of the finite system size can be used to determine some exponents, or can be taken into account by finite-size scaling. But if most of your data are made at one system size, and fewer runs are made at a drastically different system size, then a much simpler comparison of the two results tells you what order of magnitude the finite-size effects are. Quite often it also happens that programming errors give an influence which vanishes for system size going to infinity. In addition, you may get some systematic errors from random number generators which are not really random; trying different generators may, or may not, help.

In summary, it is difficult but possible to extract exponents from high quality data with an accuracy of the order of one per cent. It is much easier to estimate errors of that order by using some standard formula ignoring systematic deviations, but then it is likely that later one will regret having published such an overly optimistic error bar. My best advice is to try different methods and use some fantasy; different problems and different data qualities require different methods of analysis.

A.3. Computerized cluster counting

The preceding sections of this Appendix discussed how to analyse cluster numbers. Redner's paper mentioned after Chapter 2 gives a short program for counting exactly the number of different cluster configurations for a given size. But how do we count clusters in a Monte Carlo sample of a large lattice? And how do we check whether a cluster connects top and bottom in such a sample? If one tries to do that visually one will presumably make some errors in a large lattice. Thus we now explain how to teach a computer to do that work for us. We restrict ourselves to the algorithm of Hoshen and Kopelman since that allows the simulation of large lattices without having to store the whole lattice. Leath's entirely different algorithm which involves letting one cluster grow has already been mentioned in Chapter 2.

What we would like to have is an algorithm which gives all sites within the same cluster the same label, and gives different labels to sites belonging to different clusters. Then the top of a sample is connected to its bottom if the same

Figure 29. Illustration of a 3 × 5 square lattice with 11 occupied sites, to be analysed by the Hoshen–Kopelman algorithm.

label appears in both the top and the bottom line or plane. And by counting how many sites have the same label we get the cluster size s.

Unfortunately, life is more difficult than our dreams. Let us look at the example of Figure 29 and analyse it in the same way as you read this book: from left to right within each line, and then from the top line to the bottom line. We will give the first occupied site in the left upper corner the label 1; its neighbour to the right is empty and needs no label; then follows another occupied site, labelled by a 2 and other empty site, and finally an occupied site labelled by a 3. The next line starts with an occupied site which is a neighbour to the occupied site labelled by a 1 in the first row. Thus we label this site by a 1, too. The next site is empty; the third site is labelled by a 2 since it is directly below the occupied site labelled 2 in the first row. The fourth site is neighbour to the third one and thus also labelled by a 2. Thus when we are looking at the fifth site our labels so far are:

$$1 \quad 0 \quad 2 \quad 0 \quad 3$$

$$1 \quad 0 \quad 2 \quad 2 \quad ?$$

where we marked the empty sites by zeros. What label do we choose for the fifth site denoted here by a question mark? Its lefthand neighbour says it is a 2 whereas its top neighbour claims it as its neighbour with a 3 label. In this situation similar to that in the Fashoda swamps in 1898, one needs an *entente cordiale*. In reality all sites labelled 3 and all sites labelled 2 belong to one joint cluster which we label 2 in order to keep the label numbers as small as possible.

Having agreed on that common label at the question mark site, do we now have to go back to the beginning of the whole lattice to relabel all 3s into 2s? For small lattices we could do that and thus solve all these label conflicts by starting all over. But for large lattices that would make the computing time prohibitively large. We would like to have a computing time proportional to the number of lattice sites, and not to the square of that number, for large lattices.

Hoshen and Kopelman found a way to avoid this tedious relabelling. They follow some politicians or some good old Western movies from Hollywood by dividing the whole set of labels into the good ones and the bad ones. The good ones are those which characterize different clusters. The bad ones are those which indicate that what first looked like a new cluster later turned out to be part of an old cluster. To distinguish easily between the good and the bad guys, we cannot distribute white and black hats to them. Thus instead we introduce an additional array, the labels of labels, and denote it as N. A good label, say M, is characterized by $N(M) = M$ whereas the label N of a bad label is taken as the label to which that bad label turned out to be connected.

Thus in our case, before we came to the question mark, we regarded our three labels as good and had $N(1) = 1$ when we found the first occupied site labelled by 1, took $N(2) = 2$ when we came to the second label 2, and set $N(3) = 3$ when the third label occurred for the first time. No changes were made in the array N when further sites with labels 1 and 2 were observed. Now we are at the site with the question mark. We give that site the label 2; note that 3 turned out to be a bad label since its sites are connected with sites of label 2, and thus take $N(3) = 2$.

Now we go to the third line of our above lattice. The first site is occupied and connected to the top site; thus we label it by 1 and leave $N(1)$ unchanged. The second site is labelled by a 1, since it too is connected to its left neighbour. At the third site we have a label conflict again since labels 1 (to the left) and 2 (to the top) turn out to belong to the same cluster. The good label is the smaller one, that means 1, and the bad label is the larger one, that means 2. Thus with $N(2) = 1$, we have marked that sites with label 2 belong to the same cluster as sites with label 1. Now our labels are

$$
\begin{array}{ccccc}
1 & 0 & 2 & 0 & 3 \\
1 & 0 & 2 & 2 & 2 \\
1 & 1 & 1 & 1 & 0
\end{array}
$$

$$N(1) = 1 \qquad N(2) = 1 \qquad N(3) = 2$$

In this way we can go through the whole lattice once and store the connections found later in the 'label of labels' array N. Once finished, for any site of the lattice we find its good label by the following classification: if the label of that site is M then check first if $N(M) = M$. If yes, the label is good and we go to the next site; otherwise the label is bad and equals, say, $N(M) = M'$. Now we check whether M' is a good label, $N(M') = M'$, or a bad one, $N(M') = M''$. In the first case the number M' is the root of our label tree and classifies the given site; in the other case we check if $N(M'')$, and so on, until we come to a good label with $N(\text{label}) = \text{label}$. With $M = 3$, in our case we have $N(3) = 2$ and thus $M' = 2$; then $N(2) = 1$ and thus $M'' = 1$. Since $N(1) = 1$, M'' is a good label whereas M and M' are bad.

A simple Fortran subroutine like

```
       FUNCTION KLASS (M)
       DIMENSION N(25000)
       COMMON N
  1    MS = M
       M = N(M)
       IF(MS.NE.M) GO TO 1
       KLASS = M
       RETURN
       END
```

accomplishes the above classification: For an occupied site with original label M it searches for the good label at the root of the label tree and calls that good label KLASS.

It is practical when going through the lattice not to shift all classifications to the end of the calculation, although every site could get its good label with a second sweep through the lattice. Instead, before we assign a label to the newly investigated site, we may reclassify its two neighbours (in a square lattice), which got labels before, by the function KLASS. After the two currently good labels of these two neighbours have been found, the new site gets the smaller one as its own label, and the larger label is classified as bad and as connected to the smaller label if the site was occupied. This reclassification of the neighbours has the effect that after finishing one line (or plane in three dimensions) we can already tell if at least one occupied site is connected to the top line or plane. For if none of the labels of the line (or plane) just completed agrees with a label occurring in the first line (or plane) of the lattice, then the first line is not connected to the line under current investigation and thus also not to the bottom line. The investigation can then be stopped if we only want to know if top and bottom are connected.

Minor remarks: Since in the case of a label conflict we take the smaller one, it is practical to take a very large number, like MAX with $N(\text{MAX}) = \text{MAX}$, as the label of empty sites. Then by simply looking at the smallest of all labels assigned to the previously investigated and reclassified neighbours we get the new label; we do not have to distinguish carefully between occupied and empty sites. Also, the left neighbour needs no reclassification since its label was set in the immediately preceding step and did not have the time to become bad. Finally, time-consuming IF conditions are avoided if we add an empty column to the left boundary of our square, and perhaps also an empty line to the top boundary. A Fortran program based on these ideas (taken from unpublished work of P.J. Reynolds and J. Kertész) was published by Stauffer for the simple cubic lattice. It can easily be simplified for a square lattice or be modified for a triangular lattice. (A triangular lattice is a square lattice with site (i, j) connected to the three previously investigated sites $(i-1, j)$, $(i, j-1)$ and $(i-1, j-1)$.)

Often one wants not just to see if a lattice percolates but how many clusters of what size it contains. Then a modification is necessary. Now the bad labels have negative $N(\text{label})$, the good ones have positive $N(\text{label})$. The bad label's N equals minus the label to which it is connected (instead of plus that value in the connectivity check above), and the good label has an $N(\text{label})$ equal to the size of the cluster to which it belongs at present. Thus when a label conflict occurs, we select again as the new label the minimum of the (reclassified) labels of the previously investigated neighbours. But for N of this label we take the sum of the N of the previously separated clusters, plus unity for the site which we have just added. The classification subroutine now reads

```
      FUNCTION KLASS(M)
      DIMENSION N(25000)
      COMMON N
      MS = N(M)
      IF(MS.LT.0) GO TO 1
      KLASS = LEV
      RETURN
1     KLASS = - MS
      MS = N(KLASS)
      IF(MS.LT.0) GOTO 1
      KLASS = M
      N(M) = - KLASS
      RETURN
      END
```

(Some computers might complain that this is not really a function but a subroutine since we take the opportunity in the last statement to let the label $N(M)$ of the old label M point from now on to the just found good label KLASS.) A program for simple cubic bond percolation with this classification routine was published by Stauffer, Coniglio and Adam (1982) in the review mentioned after the Introduction.

Computer time is saved if the classification routine is not regarded as a separate function KLASS but written into the main program. We then work in the simple cubic or triangular lattice with three previously investigated neighbours ('top', 'back' and 'left') and have to reclassify only two of them (top and back) if the left neighbour was dealt with in the previous step. We now list a complete Fortran program counting clusters on a $L \times L$ triangular lattice at various concentrations p, with $L = 500$. This half-minute test run was made on a CDC Cyber 76 computer; an IBM 3081 may take about a minute.

```
      PROGRAM PERC(OUTPUT,TAPE6 = OUTPUT)
      DIMENSION LEVEL(501),N(25000),NS(18)
      LOGICAL TOP,LEFT,BACK
C     TRIANGULAR SITE L*L PERCOLATION AT CONCENTRATION P
      CALL RANSET(1)
      L = 500
      LP1 = L + 1
      LARGE = 1024
      ALOG2 = 1.0000001/ALOG(2.0)
      MAX = 25000
      MAX3 = MAX*3
      N(MAX) = MAX
      DO 7 IP = 1,25
      P = 0·37 + 0·01*IP
      DO 2 I = 1,18
2     NS(I) = 0
      INDEX = INF = 0
      CHI = 0.
      DO 1 I = 1,LP1
```

```
  1       LEVEL(I) = MAX
  C       NOW ALL INITIAL CONDITIONS ARE SET
  C       THE FIRST LINE AND LEFTMOST COLUMN IS EMPTY
          DO 3 K = 2,LP1
          IF(INDEX.GT.24700) GOTO 7
  C       DANGER OF MEMORY OVERFLOW, STOP WORKING
          MOLD = MAX
          DO 3 I = 2,LP1
          LBACK = MBACK = MOLD
          MOLD = LEVEL(I)
          IF(RANF(I).GT.P) GOTO 9
  C       JUMP TO 9 IF NEW SITE IS EMPTY
          MLEFT = LEVEL(I - 1)
          MTOP = LTOP = LEVEL(I)
          IF(MLEFT + MTOP + MBACK.EQ.MAX3) GOTO 4
  C       JUMP TO 4 IF ALL THREE NEIGHBOURS ARE EMPTY
          LEFT  = MLEFT .LT.MAX
          TOP   = MTOP  .LT.MAX
          BACK = MBACK.LT.MAX
  C       FIRST HOSHEN-KOPELMAN CLASSIFICATION OF TOP NEIGHBOUR
          IF(.NOT.TOP.OR.N(LTOP).GE.0) GOTO 12
          MS = N(LTOP)
  13      MTOP = -MS
  C       THIS IS THE FUNCTION KLASS WRITTEN INTO MAIN PROGRAM
          MS = N(MTOP)
          IF(MS.LT.0) GOTO 13
          N(LTOP) = -MTOP
  C       NOW COMES THE BACK NEIGHBOUR (LEFT OF TOP)
  12      IF(.NOT.BACK.OR.N(LBACK).GE.0) GOTO 11
          MS = N(LBACK)
  14      MBACK = -MS
          MS = N(MBACK)
          IF(MS.LT.0) GOTO 14
          N(LBACK) = -MBACK
  C       LEFT NEIGHBOUR NEEDS NO RECLASSIFICATION
  11      LEVEL(I) = MNEW = MIN0(MTOP,MBACK,MLEFT)
  C       MIN0 GIVES SMALLEST OF SEVERAL INTEGERS
          ICI = 1
          IF(TOP) ICI = ICI + N(MTOP)
          IF(LEFT.AND.MTOP.NE.MLEFT) ICI = ICI + N(MLEFT)
          IF(BACK.AND.MBACK.NE.MLEFT.AND.MBACK.NE.MTOP)
        1 ICI = ICI + N(MBACK)
          N(MNEW) = ICI
  C       ICI IS THE SIZE OF THE CLUSTER AT THIS STAGE
          IF(TOP  .AND.MTOP  .NE.MNEW) N(MTOP  ) = -MNEW
          IF(LEFT .AND.MLEFT .NE.MNEW) N(MLEFT ) = -MNEW
          IF(BACK.AND.MBACK.NE.MNEW) N(MBACK) = -MNEW
          GOTO 3
  4       LEVEL(I) = INDEX = INDEX + 1
  C       START OF NEW CLUSTER
          N(INDEX) = 1
          GOTO 3
```

```
9        LEVEL(I) = MAX
3        CONTINUE
C        NOW FINAL ANALYSIS
         IF(INDEX.EQ.0) GOTO 5
         DO 6 IS = 1,INDEX
         NIS = N(IS)
C        IF NIS < 0 IT IS A BAD LABEL AND SHOULD BE IGNORED
         IF(NIS.LT.0) GOTO 6
C        NIS IS THE NUMBER OF CLUSTERS CONTAINING S SITES EACH
C        INF IS THE SIZE OF THE LARGEST CLUSTER
C        CHI IS THE SECOND MOMENT, RELATED TO MEAN CLUSTER SIZE
         IF(INF.LT.NIS) INF = NIS
         FNIS = NIS
         CHI = CHI + FNIS*FNIS
         IF(NIS.GE.LARGE) WRITE(6,97) NIS
C        LARGE CLUSTERS ARE PRINTED OUT SEPARATELY, SMALLER
C        ONES ARE PUT TOGETHER IN BINS FROM 2**(I − 1) TO 2**I − 1
         NIS = ALOG(FNIS)*ALOG2 + 1.
         NS(NIS) = NS(NIS) + 1
6        CONTINUE
         CHI = (CHI − FLOAT(INF)**2)/(L*L)
5        WRITE(6,96) NS
96       FORMAT(" NS :",4I8,4I6,/,10I6)
         FORMAT(" CLUSTER OF SIZE ",I12)
98       FORMAT(F10·5, 3I10,F20·4,/)
7        WRITE(6,98)P,L,INDEX,INF,CHI
         STOP
         END
```

We see from this computer program that the original labels, called LEVEL here, do not require an $L \times L$ array; only one line of length L has to be stored in two dimensions, and one plane in three. However, the array N which stores the labels of labels has to have a size proportional to the number of sites in the system. We took its size here as 25 000 with $L = 500$ being our maximum size. Thus N requires about one tenth of the system size in memory. Actually, every occupied site with all previously investigated neighbours being empty increases our index for the labels by unity and thus requires one more memory element in the array N. The probability for this event to happen is $p(1 − p)^3$ in the triangular or simple cubic lattice and agrees reasonably, if multiplied by L^2, with the final value of INDEX printed out below. (For bond percolation at very small concentration p the array N must be about as large as the whole lattice, if one takes no precaution to throw out isolated sites.)

Periodic boundary conditions often diminish, sometimes enhance the influence of the lattice boundaries; in any case they make the program more complicated. Computer memory can be saved, or larger systems be simulated, if labels of labels which are no longer used in the array N are recycled like used paper. This trick is described in Chapter 8 of Binder's 1984 book on Monte Carlo methods, mentioned after our Chapter 2, with a complete program given there.

Thus we will not go into these details and now look at the output from the simpler program listed above.

In the following computer output from a CDC Cyber 76 we give the results for many different concentrations p. In every example, the first two lines give the cluster numbers in bins of exponentially increasing size: $s = 1$, $s = 2$ and 3, s from 4 to 7, then from 8 to 15, etc., as mentioned in Section 2.7. Then we list five other interesting quantities: the concentration p, the lattice size L, the maximum index needed (a number which must be smaller than the memory size reserved for the array N), the size of the largest cluster, and the second moment of the cluster size distribution (which enters the mean cluster size). Also, we list the precise size of all clusters containing at least $LARGE = 1024$ sites, in order to study the second or third largest cluster, too. Thus it already becomes optically transparent from the output that lots of large clusters only occur close to the percolation threshold. In Section 2.7 we have described how we analysed these low-quality Monte Carlo data.

15.52.50 00031·711 USR.   31·405 CP SECONDS EXECUTION TIME.

```
NS:    5493    3851    2640    1720   858    416    112    14
    0      0       0       0      0      0      0      0     0
   ·38000         500       22838        211             10·6030

NS:    4994    3458    2331    1595   878    447    149    26
    0      0       0       0      0      0      0      0     0
   ·39000         500       22094        218             13·3170

NS:    4754    3326    2162    1461   876    450    185    35
    2      0       0       0      0      0      0      0     0
   ·40000         500       21838        295             16·0111

NS:    4458    2969    1973    1309   846    468    197    52
    5      1       0       0      0      0      0      0     0
   ·41000         500       21225        614             20·1037

NS:    4124    2738    1810    1261   736    462    216    72
    9      1       0       0      0      0      0      0     0
   ·42000         500       20725        540             26·4825

NS:    3680    2447    1657    1033   671    414    200   105
   24      2       0       0      0      0      0      0     0
   ·43000         500       20009        664             38·0001
```

CLUSTER OF SIZE   1132

```
NS:    3436    2292    1349     963   607    328    190   109
   37      9       1       0      0      0      0      0     0
   ·44000         500       19381       1132            57·3412

NS:    3138    2010    1289     816   539    338    217   120
   45     10       0       0      0      0      0      0     0
   ·45000         500       18703        870             62·4797
```

CLUSTER OF SIZE   2205
CLUSTER OF SIZE   1808
CLUSTER OF SIZE   1063

NS: 2911   1856   1133   724   468   281   169   95
   51   26   2   1   0   0   0   0   0   0
  ·46000   500   18201   2205   107·4869

CLUSTER OF SIZE   1493
CLUSTER OF SIZE   2225
CLUSTER OF SIZE   1045
CLUSTER OF SIZE   2498
CLUSTER OF SIZE   2160
CLUSTER OF SIZE   2267
CLUSTER OF SIZE   1498
CLUSTER OF SIZE   1038
CLUSTER OF SIZE   1328
CLUSTER OF SIZE   1030

NS: 2541   1671   985   602   348   220   134   80
   50   31   6   4   0   0   0   0   0   0
  ·47000   500   17574   2498   206·8616

CLUSTER OF SIZE   1536
CLUSTER OF SIZE   2335
CLUSTER OF SIZE   3054
CLUSTER OF SIZE   1552
CLUSTER OF SIZE   1252
CLUSTER OF SIZE   4043
CLUSTER OF SIZE   1209
CLUSTER OF SIZE   1479
CLUSTER OF SIZE   1933
CLUSTER OF SIZE   2240
CLUSTER OF SIZE   2204
CLUSTER OF SIZE   1479
CLUSTER OF SIZE   6700
CLUSTER OF SIZE   2149
CLUSTER OF SIZE   3078
CLUSTER OF SIZE   2511
CLUSTER OF SIZE   1303
CLUSTER OF SIZE   1709
CLUSTER OF SIZE   1625
CLUSTER OF SIZE   1112
CLUSTER OF SIZE   1538
CLUSTER OF SIZE   1286

NS: 2423   1420   876   530   280   182   91   51
   36   17   13   8   1   0   0   0   0   0
  ·48000   500   16945   6700   425·9939

CLUSTER OF SIZE   9668
CLUSTER OF SIZE   2541
CLUSTER OF SIZE   1310
CLUSTER OF SIZE   1317
CLUSTER OF SIZE   1491
CLUSTER OF SIZE   2200
CLUSTER OF SIZE   2550
CLUSTER OF SIZE   3695
CLUSTER OF SIZE   1190
CLUSTER OF SIZE   2749

CLUSTER OF SIZE	10429						
CLUSTER OF SIZE	2199						
CLUSTER OF SIZE	2506						
CLUSTER OF SIZE	15059						
CLUSTER OF SIZE	1272						
CLUSTER OF SIZE	1075						
CLUSTER OF SIZE	2073						
NS: 2281	1347	797	456	250	137	77	48
22 14	3	8 0	3	0	0	0	0
·49000	500	16428	15059			1117·5676	

CLUSTER OF SIZE	19937						
CLUSTER OF SIZE	16111						
CLUSTER OF SIZE	3550						
CLUSTER OF SIZE	4311						
CLUSTER OF SIZE	24895						
CLUSTER OF SIZE	1147						
CLUSTER OF SIZE	5195						
CLUSTER OF SIZE	1876						
CLUSTER OF SIZE	2170						
CLUSTER OF SIZE	1307						
NS: 1992	1173	699	367	195	89	58	32
10 13	3	2 2	1	2	0	0	0
·50000	500	15729	24895			2947·0190	

CLUSTER OF SIZE	76513						
CLUSTER OF SIZE	1964						
CLUSTER OF SIZE	1123						
CLUSTER OF SIZE	1875						
CLUSTER OF SIZE	2339						
CLUSTER OF SIZE	10860						
CLUSTER OF SIZE	1453						
NS: 1868	1041	549	312	143	84	39	19
14 3	4	1 0	1	0	0	1	0
·51000	500	15306	76513			554·8937	

CLUSTER OF SIZE	2272						
CLUSTER OF SIZE	99041						
CLUSTER OF SIZE	1293						
CLUSTER OF SIZE	3145						
NS: 1667	951	518	254	149	54	40	13
3 3	1	2 0	0	0	0	1	0
·52000	500	14754	99041			77·0606	

CLUSTER OF SIZE	118338						
NS: 1456	818	401	218	89	33	13	7
2 0	0	0 0	0	0	0	1	0
·53000	500	14075	118338			3·0231	

CLUSTER OF SIZE	125000						
NS: 1355	663	355	144	66	18	10	2
0 1	0	0 0	0	0	0	1	0
·54000	500	13405	125000			2·0732	

CLUSTER OF SIZE   128563

NS:	1158	625	311	119	62	23	5	1
	0	0	0	0	0	0	0	1	0
	·55000	500	12959	128563				·6700

CLUSTER OF SIZE   133636

NS:	1065	526	229	101	35	11	0	1
	0	0	0	0	0	0	0	0	1
	·56000	500	12236	133636				·3457

CLUSTER OF SIZE   137726

NS:	916	436	186	87	28	8	1	0
	0	0	0	0	0	0	0	0	1
	·57000	500	11473	137726				·2070

CLUSTER OF SIZE   141958

NS:	814	336	158	60	12	1	1	0
	0	0	0	0	0	0	0	0	1
	·58000	500	10872	141958				·1042

CLUSTER OF SIZE   143765

NS:	774	325	136	47	13	5	0	0
	0	0	0	0	0	0	0	0	1
	·59000	500	10491	143765				·1177

CLUSTER OF SIZE   147201

NS:	709	266	120	26	8	1	1	0
	0	0	0	0	0	0	0	0	1
	·60000	500	10012	147201				·0963

CLUSTER OF SIZE   150802

NS:	576	241	75	14	7	1	1 Q·	0
	0	0	0	0	0	0	0	0	1
	·61000	500	9037	150802				·0617

CLUSTER OF SIZE   153402

NS:	501	217	75	24	4	1	0	0
	0	0	0	0	0	0	0	0	1
	·62000	500	8659	153402				·0461

Output of this type was used by Margolina *et al.* (1983) (cited in Chapter 2) to analyse lattices with up to 10^{10} sites, using only the core memory of a CDC Cyber 76. Clearly that would have been impossible without the Hoshen–Kopelman algorithm with its elegant classification routine and the recycling of labels of labels no longer needed in the array *N*. With additional tricks, Rapaport (1985) (as cited in Chapter 2) simulated a $160\,000 \times 160\,000$ square lattice, probably the largest system simulated so far (February 1985).

Readers interested in diffusion on disordered lattices will find Fortran programs, including those for CDC Cyber 205 vector computers, in the paper by Pandey *et al.* (1984) cited in Chapter 4. At present I do not know an efficient way to put the above cluster counting program on vector computers.

Further reading

Hoshen, J. and Kopelman, R., *Phys. Rev. B*, **14**, 3428, (1976).

Stauffer, D., page 9 in: *Disordered Systems and Localization*, edited by C. Castellani, C. di Castro and L. Peliti, (Heidelberg: Springer Verlag, 1981).

INDEX